Springer Series on Polymer and Composite Materials

Series editor

Susheel Kalia, Dehradun, India

More information about this series at http://www.springer.com/series/13173

Shivani Bhardwaj Mishra
Ajay Kumar Mishra
Editors

Bio- and Nanosorbents from Natural Resources

Editors
Shivani Bhardwaj Mishra
Nanotechnology and Water Sustainability
 Research Unit
University of South Africa
Johannesburg
South Africa

Ajay Kumar Mishra
Nanotechnology and Water Sustainability
 Research Unit
University of South Africa
Johannesburg
South Africa

ISSN 2364-1878   ISSN 2364-1886 (electronic)
Springer Series on Polymer and Composite Materials
ISBN 978-3-319-68707-0   ISBN 978-3-319-68708-7 (eBook)
https://doi.org/10.1007/978-3-319-68708-7

Library of Congress Control Number: 2017954474

© Springer International Publishing AG 2018
This work is subject to copyright. All rights are reserved by the Publisher, whether the whole or part of the material is concerned, specifically the rights of translation, reprinting, reuse of illustrations, recitation, broadcasting, reproduction on microfilms or in any other physical way, and transmission or information storage and retrieval, electronic adaptation, computer software, or by similar or dissimilar methodology now known or hereafter developed.
The use of general descriptive names, registered names, trademarks, service marks, etc. in this publication does not imply, even in the absence of a specific statement, that such names are exempt from the relevant protective laws and regulations and therefore free for general use.
The publisher, the authors and the editors are safe to assume that the advice and information in this book are believed to be true and accurate at the date of publication. Neither the publisher nor the authors or the editors give a warranty, express or implied, with respect to the material contained herein or for any errors or omissions that may have been made. The publisher remains neutral with regard to jurisdictional claims in published maps and institutional affiliations.

Printed on acid-free paper

This Springer imprint is published by Springer Nature
The registered company is Springer International Publishing AG
The registered company address is: Gewerbestrasse 11, 6330 Cham, Switzerland

Preface

Bio- and nanosorbents have been materials of interests for the researchers who believe in sustainable solutions. With rapidly growing industries using new and novel smart materials for commercial applications, the research is emphasized onto exploring ecofriendly materials to combat the challenge of pollution in all forms and to deliver the products with finest properties. Nanotechnology has played a huge role in the field of science, engineering, and technology, and the ideas have been out in the market for public utility.

Sorption and adsorption are an energy-efficient technique applicable for many important industrial and environmental applications. Adsorbents are designed based on its functional properties and feasibility to be used for large-scale processes and methodologies. Many scientific techniques use adsorption as a principle technique such as chromatography used for separation systems, sensing devices those are primarily made of composite or nanocomposite adsorbing the biofluids and providing accurate response, the material adsorbing light from visible spectrum to be used for electronic and optical devices, catalytic conversion of the petroleum products using heterogeneous catalysts, the filters used for water treatment use adsorption technique, bacterial filtration. These were few well-known examples where bio- and nanoadsorbents have strong foundation and its utilization is visible in our daily lives.

As mentioned above, the applicability of adsorbents is unique and diverse. However, the book is consolidated on the application of these sorbents in nanosize derived from natural sources to be used for decontamination of organic and inorganic pollutants. There are eight chapters dealing with the theme of various adsorbents used for wastewater treatment. Chapter "Biosorbents from Agricultural By-products: Updates After 2000s" provides a review on the update after the year 2000 on the biosorbents derived from agriculture by-products. Chapter "Carbon Nanoadsorbents for Removal of Organic Contaminants from Water" deals with the carbon nanosorbents used for the removal of organic contaminants. Chapter "Lignin and Chitosan-Based Materials for Dye and Metal Ion Remediation in Aqueous Systems" discusses the lignin which is cementing agent of plant fibers and chitosan-based nanocomposite material for the removal of organic dye and heavy metal ions. Cationic polyelectrolytes have been versatile materials for environmental cleanup which has dealt in Chapter "Cationic Nanosorbents Biopolymers: Versatile Materials for Environmental Cleanup," whereas Chapter "Alginate-Based Nanosorbents for Water Remediation" provides information about alginate-based nanosorbents for water remediation. Chapter "Chitosan-Based Natural Biosorbents: Novel Search for Water and Wastewater Desalination and Heavy Metal Detoxification" exclusively talks about chitosan polymer and its application for desalination and detoxification. Chapter "Application of Biomaterials for Elimination of Damaging Contaminants from Aqueous Media" explains the application of biological adsorbent material for removal of inorganic contaminants from aqueous media. Chapter "Synthesis and Application of Silica Nanoparticles-Based Biohybrid Sorbents" deals with nanoadsorbents for the removal of hexavalent chromium from water and wastewater.

The present book will benefit the researchers working in the area of nanoadsorbents and biosorbents used for wastewater research. These adsorbents have been utilized in its pristine form, as well as developed as polymer blends, composites, nanocomposites, and bionanocomposites. The unique properties of these sorbents lie in the fact with respect to adsorption efficiency and environmental sustainability that has been a talk of the research world and environmental protection agencies.

The book is an inspiration to the researchers working in the field of environmental protection, water engineers, and environmental managers and to all the young researchers who wish to develop their career in this field. The book presents the recent progress and future prospects that can be made in the area of bio- and nanosorbents.

Johannesburg, South Africa   Shivani Bhardwaj Mishra Ph.D., FRSC
Ajay Kumar Mishra Ph.D., FRSC

Contents

Biosorbents from Agricultural By-products: Updates After 2000s 1
Rekha Sharma, Sapna, Ankita Dhillon and Dinesh Kumar

Carbon Nanoadsorbents for Removal of Organic Contaminants from
Water. 21
Fernando Machado Machado and Éder Cláudio Lima

Lignin and Chitosan-Based Materials for Dye and Metal Ion
Remediation in Aqueous Systems 55
Thato Masilompane, Nhamo Chaukura, Ajay K. Mishra, Shivani B. Mishra
and Bhekie B. Mamba

Cationic Nanosorbents Biopolymers: Versatile Materials for
Environmental Cleanup 75
Sandeep K. Shukla, Rashmi Choubey and A.K. Bajpai

Alginate-Based Nanosorbents for Water Remediation 103
A.K. Bajpai, Priyanka Agrawal, Sunil K. Singh and Priyanka Singh

Chitosan-Based Natural Biosorbents: Novel Search for Water and
Wastewater Desalination and Heavy Metal Detoxification 123
Ankita Dhillon and Dinesh Kumar

Application of Biomaterials for Elimination of Damaging
Contaminants from Aqueous Media 145
Vaishali Tomar and Dinesh Kumar

Synthesis and Application of Silica Nanoparticles-Based Biohybrid
Sorbents 161
Ritu Painuli, Sapna Raghav and Dinesh Kumar

Biosorbents from Agricultural By-products: Updates After 2000s

Rekha Sharma, Sapna, Ankita Dhillon and Dinesh Kumar

Abstract Biosorption is the characteristic method for the binding and removal of heavy metal ions contaminants. The biosorption method is effective as an ion exchange treatment even at low metal concentration. This method has green and affordable properties in comparison to predictable techniques. The detection of metal ion toxicants occurs because of the binding of metal ions with biosorbent using biosorption technique. The purpose of this chapter is to discuss the important features of the biosorbent particularly agriculture products. The present chapter also focuses on the removal of metal ion pollutants as well as regeneration methods of the biosorbents. The chapter will help the readers to select suitable sorbents and to take up further research required for pollutant removal using biosorbents, depending on the characteristics of effluents to be treated, industrial applicability, release standards, cost-effectiveness, regulatory requirements, and durable ecological impacts.

Keywords Biosorption · Treatment · Detection · Efficiency · Cost-effective

1 Introduction

In most recent times, owing to the anthropogenic actions, expansion of population, unexpected urbanization, fast industrialization, and untrained use of natural water assets the water quality has rigorously deteriorated worldwide. The water pollution occurs mainly due to unprocessed sanitary discharge, and poisonous manufacturing squanders the discarding of manufacturing pollutants, and overflow from farm fields. Various toxic chemicals such as micro pollutants, insect killers, and toxic metal ion in the aqueous system have been found in drinking water as dangerous

---

R. Sharma · Sapna · A. Dhillon · D. Kumar (✉)
Department of Chemistry, Banasthali University, Tonk 304022, Rajasthan, India
e-mail: dinesh.kumar@cug.ac.in

R. Sharma · Sapna · A. Dhillon · D. Kumar
School of Chemical Sciences, Central University of Gujarat, Gandhinagar 382030, India

© Springer International Publishing AG 2018
S. Bhardwaj Mishra and A.K. Mishra (eds.), *Bio- and Nanosorbents from Natural Resources*, Springer Series on Polymer and Composite Materials, https://doi.org/10.1007/978-3-319-68708-7_1

effluents for water pollution. These toxicants cause various health problems to living beings. Hence, it is vital to develop vigorous, cost-effective, and ecological welcoming methods for the removal of toxic pollutants from aqueous solutions. To minimize the water contamination, there are various treatment methods utilized successfully [1]. However, in the water treatment process, most of the techniques have high operational and maintenance costs, generation of toxic sludge, and complicated procedure. Among them, the adsorption process is considered as an enhanced substitute in water treatment process due to handiness, the simplicity of operation, and ease of constructing [2]. Additionally, these processes can be able to remove diverse sort of effluents and have an extensive use in water pollution control. For the treatment of effluents and removal of various pollutants from water, activated carbon is considered as an universal adsorbent. Nevertheless, because of the higher cost of this activated carbon, its worldwide use is sometimes restricted. For the removal of different sorts of pollutants from water and wastewater, a large variety of inexpensive adsorbents has been observed [3, 4]. A low-cost adsorbent is the adsorbent which requires small processing, high abundance in the environment, and found as a by-product or waste material from the industry. For the removal of water pollutants, different adsorbents of various origins present poor adsorption potential than commercial activated carbon. Consequently, for the removal of water pollutants, the development of efficient sorbents is still in progress. There are several methods which have been studied since last few years to synthesize effective adsorbents with low-cost containing natural biopolymers. For the elimination of different kind of contaminants from the water, a variety of biopolymer materials has been observed. Owing to the exclusive chemical nature and their ease of availability and renewable nature, low-cost agricultural waste materials are economical and biodegradable, which are used for the removal of effluents from the water. Lignin and cellulose are the major components in the development of agricultural by-products. A variety of other polar functional groups of lignin also utilized in the formation of these adsorbents such as alcohols, aldehydes, ketones, carboxylic, phenolic, and ether groups. Different binding mechanisms were used to bind aquatic pollutants with these functional groups. Nowadays, water pollution is one of the significant ecological harm, because of the poisonous character and accumulation of heavy metal ion pollutants in the food chain due to their non-biodegradable nature [5–7]. The metal plating, mining operations, surface finishing industry, tanneries, paper and pulp industries, color-alkali, fertilizer and pesticide industry, radiator manufacturing, smelting, energy and fuel production, aerospace and atomic energy installation, alloy industries, electroplating, and battery industries are the various industries which cause heavy metal contamination in wastewater [8–10]. Due to the high exposure to lead, various diseases are occurring such as encephalopathy, cognitive impairment, behavioral disturbances, kidney damage, anemia, and toxicity to the reproductive system [11]. The hexavalent form of chromium (Cr) causes lethal effects owing to its strong oxidation properties [12, 13]. The effect of Cr(VI) compounds on human being appears as a higher incidence of respiratory cancer [14, 15]. The Cr(VI) compounds reach the bloodstream, the liver, and blood cells through oxidation reactions, and causes kidney damages, irritation, and skin corrosion [16]. Cadmium (Cd) is known as a worldwide contaminant for water pollution. The most toxic "Big Three" category of heavy metals contains the cadmium, lead, and mercury, heavy metal ions due to its acute toxicity [17]. Cadmium poisoning causes an illness known as "Itai-Itai" in Japan related to consequential to several ruptures arising from osteomalacia. Various physical illnesses occur due to high dose of copper (Cu) concentrations such as weakness, lethargy, anorexia, and damage to the gastrointestinal tract [18]. Skin allergies, lung fibrosis, and cancer of the respiratory tract occur due to the presence nickel (Ni) in the human body [19, 20]. The nickel-induced carcinogenesis focuses on various epidemiologic and experimental investigations, but its exact mechanisms are not known. Various methods have been utilized for the removal of heavy metal ions from water, which is made of different techniques and technologies such as physical, chemical, and biological. For the removal of heavy metal ions from water, various methods have been used, for example, chemical precipitation [21], filtration [22], ion exchange [23], electrochemical treatment [24], membrane technologies [25], floatation [26, 27] adsorption on activated carbon [28, 29] evaporation, and photocatalysis [30]. The chemical precipitation and electrochemical treatment are ineffective and produce a large quantity of sludge and require further treatment [31]. The ion exchange, membrane technologies, and activated carbon adsorption process are extremely expensive, which are utilized in the treatment of water pollution. For the past few years, numerous novel methods have been examined for heavy metal ion removal successfully [32–34].

The present chapter focuses that various aquatic pollutants such as metals, dyes, phenolic compounds, inorganic anions, radionuclides can be removed by utilizing different agricultural peels, e.g., fruits and vegetables as an adsorbent.

2 Background of Agricultural Products

Various parts of the plants such as bark, stem, leaves, root, flower, fruit biomass, husk, hull, skin, shell, bran, and stone can be agricultural waste-based biosorbents (AWBs). Cellulose, hemicellulose and lignin from AWBs which have a high content of hydroxyl groups and their structure are shown in Fig. 1. As a result, these agricultural products have a good affinity to adsorb heavy metal ions [35]. In addition, various functional groups have also been utilized such as acetamido, carboxyl, phenolic, structural polysaccharides, amido, amino, sulphydryl carboxyl groups, alcohols, and ester form these AWBs materials. The metal ions bind with these functional groups by substitution of hydrogen ions in solution or donation of an electron pair to form complexes with the metal ions in solutions. For the cleanup of heavy metals from wastewater, AWBs are the most probable adsorbent materials due to abundant binding groups.

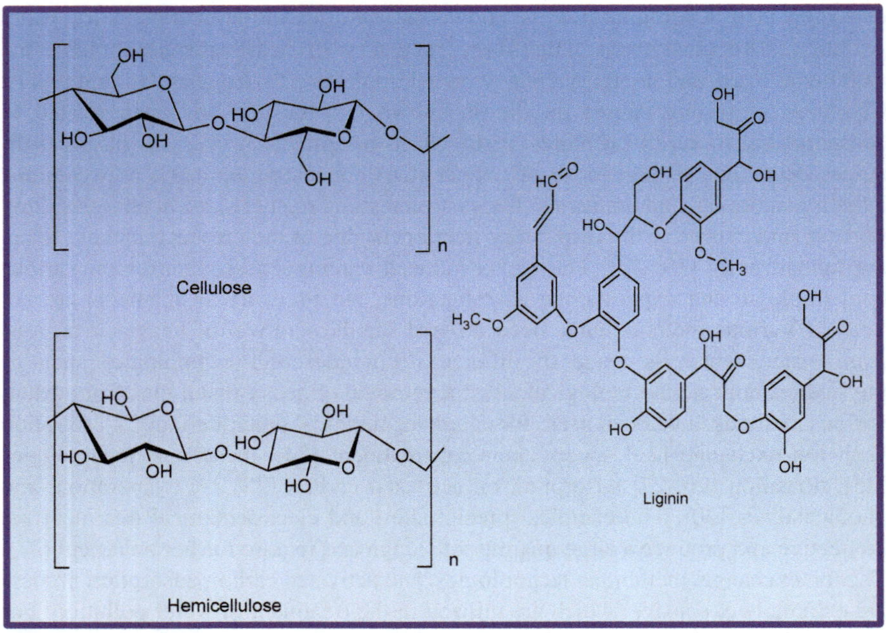

Fig. 1 Structure of cellulose, hemicellulose, and lignin

3 Agricultural Products and by-Products as Low-Cost Adsorbents

Hong and Wang applied response surface methodology for the removal of reactive dye blue 4 (RB4) using rice bran low-cost adsorbent [36]. The response surface methodology was utilized to study the effect of various parameters, and it was found that maximum adsorption capacity of the modified rice bran adsorbent was 151.3 mg/g at the dye concentration of 500 ppm, adsorbent dosage of 65.36 mg, and temperature of 60 °C. The obtained maximum adsorption capacity was comparable with other agricultural waste adsorbents reported in the literature. The endothermic and spontaneous monolayer adsorption of RB4 was confirmed demonstrated by thermodynamics analysis studies [36]. The fly ash-based adsorbent material-supported zero-valent iron was developed by Liu et al. for the removal of Pb(II) and Cr(VI) from aqueous solutions [37]. The developed adsorbent contained fly ash as skeletal material, bentonite as a binder and Enteromorpha proliferate as pore former. The coke was used as a reductant for the direct reduction of iron ore tailings to develop zero-valent iron. SEM analysis showed the presence of different size pores in developed adsorbent. The incidence of iron in the zero-valent state was confirmed by XRD analysis. The FTIR analysis demonstrated the presence of various surface functional groups responsible for heavy metals uptake. The adsorbent presented higher maximum removal capacity for Pb(II) (78.13 mg/g) than Cr(VI) 15.70 mg/g. The low cost, high efficiency, and reusability of the adsorbent make it a promising technique for heavy metal adsorption [37]. Kılıc et al. utilized reported biochar, a by-product of pyrolysis of the almond shell for the adsorption of Ni(II) and Co(II) heavy metal ions [38]. Magnetic biosorbent based on oil palm empty fruit bunch fibers, cellulose, and Ceiba pentandra was utilized by Daneshfozoun et al. for Pb(II), Cu(II), Zn(II), Mn(II), and Ni(II) ions uptake [39]. For the first time, the authors developed agro-based magnetic biosorbents (AMBs) from Ceiba pentandra (RKF), oil palm empty fruit bunches (EFB), and celluloses (CEL) extracted from EFB. The adsorbent showed optimum sorption between pH 5–7. Further, increase in both initial ion concentration and solution temperature increased the rate of sorption process. The recyclability of AMBs was checked up to 5 adsorption/desorption cycles. The magnetic biosorbent based on kapok had maximum Pb(II) removal efficiency of 99.4% than cellulose (98.2%) and EFB (97.7%). Further, as compared to raw sorbents, the magnetic sorbents exhibited higher adsorption efficiency [39]. Wu et al. carried out fabrication of carboxymethyl chitosan-hemicellulose resin (CMCH) by a thermal crosslinking process for adsorptive removal of Ni(II), Cd(II), Cu(II), Hg(II), Mn(VII), and Cr(VI) metals from wastewater [40]. The kinetics studies followed pseudo-second-order model and adsorption isotherm studies followed Langmuir model. Regeneration studies presented CMCH as a potentially recyclable adsorbent for metal ions uptake [40]. Coconut husk, i.e., unmodified coconut (Cocos nucifera L.) husk was utilized by Malik et al. for the removal of Ni(II), Pb(II), Cu(II), Zn(II) [41]. The experimental studies presented maximum adsorption at 443.0 mg/g (83.6%) for Cu(II), for Ni(II) with 404.5 mg/g (80.9%), 362.2 mg/g (72.4%) for Pb(II), and 338.0 mg/g (67.6%) for Zn(II) ion. Quantum-based theoretical study presented the rich electron donation sites of coconut husk as the main sites for the metal ions uptake [41]. Shakoor and Nasar utilized *cucumis sativus* peel waste as a low-cost adsorbent adsorptive treatment of hazardous methylene blue dye from artificially contaminated water [42]. The heterogeneous nature of the adsorbent surface was supported by Freundlich adsorption isotherm. Further, kinetics studies supported pseudo-second-order kinetics. The process was found to be spontaneous and exothermic in nature as demonstrated by thermodynamic studies. The highest regeneration of adsorbent was achieved using hydrochloric acid [42].

4 Agro Sorbents

For the removal of heavy metal ions, Schiewer and Patil [43] discovered a biosorbents made of pectin-rich fruit wastes (Fig. 2). The studies demonstrated pectin-rich fruit materials and citrus peels found as most suitable biosorbent for the biosorption of cadmium. The kinetics was favored using a first-order model. The adsorption isotherm well fitted to Langmuir model. On decreasing pH, the metal uptake was decreased, demonstrating the competition of protons for binding to acidic sites.

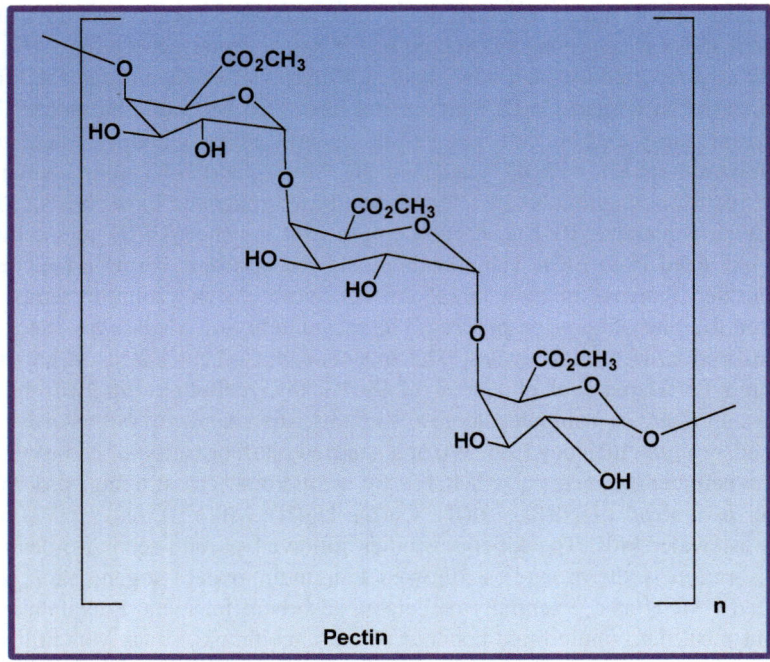

Fig. 2 Structure of pectin

For the removal of heavy metals such as Cr, Ni, and Cd, the adsorption potential of agricultural residues such as rice straw, wheat straw, and Salvinia plant biomass have been studied by performing batch adsorption experiments. The removal efficiency was increased at low metal concentration (35 mg/L) of toxic metal ions. Salvinia biomass showed advanced efficiency among the combination of three materials, i.e., rice straw, wheat straw, and Salvinia biomass which were used for the removal of heavy metals such as Cr, Ni and Cd. El-Said et al. [44] demonstrated the adsorption of Zn(II) and Se(IV) from their aqueous solution using rice husk ash, an agricultural waste, as an adsorbent. Various parameters have been studied such as contact time, metal ion concentration, adsorbent dose, and pH at 258 °C. The Zn(II) ions have higher and faster adsorption capability and adsorption rate than for Se(IV) ions. The equilibrium time for the removal of Zn(II), and Se(IV) was found within 1 h and 100 h, respectively. On increasing biosorbent dosage from 1 to 10 g/L and diminishing initial metal ions concentration, the adsorption of metal ions was increased. The Zn(II) ions showed higher affinity than that for Se(IV) ions by Rice husk ash biosorbent. For the removal of cadmium ions from aqueous solution, Ho and Ofomaja [45] used a waste product known as coconut copra for the synthesis of biosorbent. The adsorption data well fitted to three widely used isotherms, Langmuir, Freundlich, and Redlich–Peterson and examined a comparison of the linear least-squares method and a trial and error nonlinear method of these isotherms. In another study, the removal of Cd(II) metal was investigated using pectin by Schiewer and Iqbal [46] in citrus peels, native orange peels, protonated peels, depectinated peels, and extracted pectic acid. The equilibrium was accomplished within 1 h. The potentiometric titrations show that carboxyl groups of pectin were a major contributor to the surface charge of the negative surface charge. The optimized pH for metal binding experiments was obtained at pH 5. The adsorption data well fitted to the Langmuir isotherm model. The first-order model better described the metal binding kinetics than by the second-order model. The removal of Cr(III), Fe(III), and Zn(II) metal ions from aqueous solutions were reported by Vaghetti et al. [47] by checking the feasibility of pecan nutshell (Caryaillinoensis) as a biosorbent. Various parameters have been studied on the adsorption capacities of pecan nutshell, such as pH and the biosorbent dosage. The adsorption data were best fitted to Sips isotherm models according to statistical error function. For the removal of Cr(III), Fe(III), and Zn(II) metal ions, the maximum biosorption capacity of pecan nutshell was 561 93.01, 76.59, and 107.9 mg/g, respectively.

5 Lignocellulosic Wastes and by-Products for Heavy Metal Ion Removal

Lignocellulosic waste materials have proper characteristics and structural compounds to adsorb heavy metal ions on their surface binding sites through interaction with the chemical functional groups. Due to the different laboratory conditions (e.g., pH, temperature, adsorbent dose, particle size), materials and methodologies, it is not easy to conclude which biosorbent would be suitable for metal ion. For example, sugarcane bagasse, an agro-waste from sugar industries has been extensively studied because of its low price and high availability all over the world. This biosorbent exhibited very high potential in heavy metal uptake during wastewater treatment [48, 49]. The biosorption capacity of sugarcane bagasse could be noticeably improved by introducing carboxylic, amine and other functional groups into the surface materials [50] or by removing soluble organic compounds, and increasing efficiency of metal sorption [51]. The main factors determining the adsorption of As(V) on sugarcane bagasse modified by Fe(III) oxy-hydroxide (HFO) were electrostatic interactions, ligand exchange, and chelation between positively charged surface groups of FeOH (II), and negatively charged As(V) ions [52]. In another study, Krishnani et al. reported biosorption mechanism of nine different heavy metals onto bio-matrix from rice husk. In this study, rice husk was subjected to 1.5% alkali treatment and used for zinc, copper, cadmium, nickel, lead, manganese, cobalt, mercury, and chromium removal from single ion and mixed solutions. The Langmuir isotherm model well fitted to adsorption isotherm. The order of biosorption efficiency was found as Ni(II) < Zn(II) < Cd(II) < Mn(II) < Co(II) < Cu(II) < Hg(II) < Pb(II) at 32 ± 0.5 between pH range of 5.5 and 6.0 ± 0.1 [53]. However, in other study, the pretreatment of sugarcane bagasse using NaOH and HCl had no significant effect on mercury biosorption capacity [54]. The other typical and abundantly available agro-industrial materials are wheat and barley wastes, as the main crops all over the world. Pehlivan et al. reported the adsorption of copper, using esterified barley straw which was thermochemically modified with citric acid [55]. Increasing the temperature, improved the reaction efficiency, but led to lower carboxyl content and increased crosslinking of modified barley straw. Increasing citric acid concentration enhanced free carboxyl groups on the biomass matrix. Further, a significant amount of free –COOH groups remained in the biomass structure up to 4 h and then the increase in crosslinking occurs with time. Miretzky and Cirelli [56] reported that alkali treatments in comparison with acidic ones at the same conditions were more effective on metal ion removal by solving the cell wall matrix. Therefore, alkali treatments could result in better diffusion through walls and make the functional groups densest and thermodynamically more stable. The increase in Cd(II) and Cu(II) uptake on wheat straw with temperature was attributed to the increased available active sites on the surface of the adsorbent as a result of the opening of the cellulose fibers of wheat straw in a warmer solution [57]. The adsorption study on sawdust has been reported by Asadi et al. [58], Bulut and Tez [59], and Prado et al. [60]. It has been identified that phenolic, hydroxyl, and carboxylic functional groups of sawdust are responsible for heavy metal uptake, as heavy metal ions could accumulate in the secondary septum of wood in which the amount of lignin is very low. Palumbo et al. [61] found that metal adsorption onto sorbent is possibly a passive binding. They highlighted the effect of pH and natural organic matter on Zn removal. Šciban et al. [62] investigated heavy metal biosorption capacity of sawdust in synthetic and real cable factory wastewater treatment. Heavy metal ion adsorption was influenced by the existence of other ions and organic materials in real wastewater through three phenomena, namely, synergism, antagonism, and non-interaction. This study showed that copper was better adsorbed from wastewater with multiple heavy metal ions from wastewater containing single Cu(II) ion. Whereas, Cd(III) adsorption was inhibited by other metal ions and Zn(II) removal were unaffected. As expected, some metal ions have better affinity than other ions towards lignocellulosic biosorbents due to the selectivity potential of functional groups. The experimental data of Brunauer—Emmett—Teller (BET), Langmuir and Freundlich models fitted for the removal of these heavy metal ions. Bulut and Tez reported that Ni, Cd, and Pb biosorption on walnut tree sawdust was favorable at higher temperatures. The randomness at solid–solution interface increased by temperature leading to enhanced adsorption at higher temperatures. This may be related to adsorption surface activation and/or pore size enlargement. Reverse trend was obtained for Pb biosorption on different types of sawdust as an exothermic process. As a pectin-rich by-product of fruit juice industry, the suitable chemical treatment [e.g., mercapto-acetic acid ($C_2H_4O_2S$) and carbon disulfide (CS_2)] can make orange peels more favorable for metal adsorption due to the negativity amount of zeta potential, indicating higher physical stability and surface activity [63]. In another study, the biosorption properties and mechanical strength of the olive stone modified by H_2SO_4 (1 M) have been discussed for lead removal in a fixed-bed in column. The column was retained over 14 cycles of use, and life factor revealed that biosorbent bed would be exhausted after 71.3 cycles. Furthermore, according to the work study mentioned above, all biosorption was found to achieve equilibrium in a very short contact time and all kinetic studies showed the applicability of pseudo-second-order kinetic model and the second-order nature of biosorption process of heavy metal ions onto raw and pretreated biosorbents. This can be attributed to assumption of chemical adsorption rate-controlling step in biosorption process involving electron sharing or transferring between adsorbent and adsorbate.

6 Biosorption Process

Biosorption is a widely utilized method for the remediation of various water contaminants. Since the last decade, numerous biomasses of different genre have been known to have good biosorption capacity. For the development of basic scientific understanding of the biosorption process various biosorption mechanisms have been proposed by researchers. Yet, large-scale commercial applications of biosorption method need to be explored widely. Some of the important factors that govern the growth and development of biosorption as a practical technique for wastewater decontamination are, (1) further requirement of investigations on multi-component solutions, (2) lack of full understanding of physicochemical properties of biomasses of different origin, and (3) requirement of improved performance of biosorbents via surface functionalization. Therefore, a critical evaluation of the practical limitations of biosorption and future research directions are required to develop biosorption a technologically towards desired treatment applications and large-scale applications [64]. Biosorption is a process of passive uptake of contaminants using dead or inactive biological materials via various physicochemical mechanisms like physical adsorption, ion exchange, chelation, complexation, and micro-precipitation [65, 66]. Algae (fresh and marine), fungi, bacteria, industrial wastes, agricultural wastes, and other polysaccharide materials are the most commonly employed biosorbents for the heavy metal ions uptake [67–69].

7 Applications of Biosorbents

The applications of biosorbents for the removal of heavy metal ions and other organic and inorganic micro pollutants are known all around the world. Haq et al. carried out biosorption of Pb(II) and Co(II) from aqueous solutions potential using Rosa gruss an teplitz (Red Rose) Waste Biomass (RRWB) [70]. Experimental equilibrium studies showed that the biosorption capacity of Co(II) (115.9 mg/g) was higher than Pb(II) (112.0 mg/g). Equilibrium kinetics studies demonstrated pseudo-second-order kinetics. Further, pretreatment with different reagents for the modification of red rose waste biomass considerably enhanced the biosorption capacity of Pb(II) and Co(II) heavy metals from aqueous solutions [70]. Similarly, Rosa bourbonia waste phyto-biomass (RBWPB) pretreated with organic acids was utilized by Manzoor et al. for the sorption of Pb(II) and Cu(II) from aqueous media [71]. It was found from the study that pretreatments of developed biosorbent with organic acids diminished the metal adsorption capacity of RBWPB. The biosorption kinetics of Pb(II) and Cu(II) was best described by pseudo-second-order kinetic. Further, thermodynamic studies supported the spontaneous nature of adsorption process onto RBWPB [71]. Orange peel was modified by Feng et al. using with sodium hydroxide and calcium chloride in order to study the adsorption behavior of Cu(II), Pb(II), and Zn(II) on modified orange peel (SCOP) [72]. Adsorption isotherm studies followed Langmuir isotherm with the maximum adsorption capacities for Cu(II), Pb(II), and Zn(II) of 70.73, 209.8, and 56.18 mg/g, correspondingly. Column studies were carried out and breakthrough points were obtained using a column packed with SCOP. Ion exchange-based mechanism involving neutralization of carboxyl groups of the pectin by Ca(II) ions [72]. In a study, pure and thioglycolic acid-treated cassava tuber bark wastes (CTBW) were investigated for the kinetics of Cd(II), Cu(II), and Zn(II) adsorption [73]. Equilibrium kinetic studies presented exothermic nature of adsorption process with physisorption as the rate limiting sorption step. Further, the kinetic studies also supported the fast and stable nature of sorption process. The Langmuir monolayer sorption capacity was in the range of 5.88–26.3, 33.3–90.9, and 22.2–83.3 mg/g for Cd(II), Cu(II), and Zn(II), correspondingly. Thermodynamic studies indicated that the spontaneous and exothermic nature of adsorption process [73]. Mishra et al. carried out the biosorption of Zn(II) using Cedrus deodara sawdust (CDS) [74]. The adsorbent had more porous, non-crystalline, and heterogeneous surface prior to the sorption of Zn(II) ions. The adsorbent presented maximum uptake adsorption capacity of 97.39 mg/g at pH 5, temperature 45 °C, the initial concentration of Zn (II) ion 100 mg/L, contact time of 150 min, and agitation rate 160 rpm. Equilibrium kinetic studies favored chemosorptive forces of attraction for the sorption of Zn(II) ion on the surface of CDS. Bangham's equation and film diffusivity showed comparatively higher sorption rate of Zn(II) ion in the early phase of contact time whereas, at the later stage of contact time intraparticle or pore diffusion of Zn(II) ion inside the pores of CDS was found to be the rate limiting step. [74]. Senthil et al. utilized sulfuric acid-treated cashew nut shell (STCNS) for the adsorption of metal ions such as Cu(II), Cd(II), Zn(II), and Ni(II) from aqueous solution [75]. The adsorption isotherm studies followed Langmuir adsorption isotherm with maximum adsorption capacity of 406.6 mg/g for Cu(II), 436.7 mg/g for Cd(II), 455.7 mg/g for Zn(II), and 456.3 mg/g for Ni(II). The spontaneous and exothermic nature of adsorption of metal ions onto the STCNS was supported by thermodynamic studies. At the earlier stages, external mass transfer of metal ions and at the later stages intraparticle diffusion controlled the rate adsorption. [75]. Farooq et al. utilized an easily available biosorbent material, i.e., powdered straw from Triticum aestivum (WS) for Cd(II) ions uptake [76]. The authors modified the straw with urea by using microwave radiation which increased the surface area of the adsorbent. Hydroxyl, carboxyl, and amino functional groups were found to be involved in the adsorption of Cd(II) ions uptake [76]. Zafar et al. utilized acid-treated (H_3PO_4) rice bran for the biosorption of nickel from aqueous medium [77]. Experimental studies showed more resistance to pH variation within the tested pH range (pH 1–7), retaining up to 102 mg/g of the nickel binding capacity at pH 6 [77]. Flores–Garnica et al. carried out batch sorption studies of Ni(II) ions using Litchi chinensis seeds (LCS) in terms of kinetics, equilibrium, and thermodynamics [78]. Maximum biosorption of 66.62 mg/g occurred at 7.5 which significantly increased with initial metal concentration and temperature. Although the adsorbents discussed above are eco-friendly in nature, yet a higher sorption capacity is required to develop their field application [78].

Therefore, it is vital to develop efficient sorbents for the removal of these pollutants. Table 1 summarizes various sorbents which are effectively removing various toxic pollutants mainly heavy metal ions present in aqueous systems.

8 Regeneration Studies

Good adsorbents have recoverability as one the properties for their commercial utilization. The regenerated adsorbent with high efficiency makes the process far more efficient and reasonably priced. Moreover, the dumping of used up adsorbent also generates a secondary source of a pollutant which can increase toxicity and health risks. Therefore, the utilization of adsorbent having regeneration ability also presents advantages of reduced secondary pollutant generation in addition to the revival of expensive adsorbate.

Ofomaja et al. treated pine cone powder surface with potassium hydroxide and utilized it for copper(II) and lead(II) removal from solution [80]. Desorption experiments were suited for the industrial applicability of developed adsorbent. Different solvents were utilized for extracting biosorbed metal ions. It has been found that if desorption occurred by water, then the metal ion is attached to the biosorbent is by weak bonds. And if it occurred by strong acids such as HNO_3 or HCl then the metal ion is attached to the biosorbent is by ion exchange [80]. In another study, biosorption of copper (Cu), zinc (Zn), and lead (Pb) on watermelon rind was investigated by Liu et al. [83]. Desorption experiments were conducted for the repeated reusability of the developed biosorbent. For that, the authors treated the metal-loaded biosorbent with various eluant solutions like distilled water, 0.1 mol/L NaOH, 0.5 mol/L HNO_3, and 0.5 mol/L HCl for 10 h. The regenerated biosorbent was then found to be applicable for next three rounds of biosorption–desorption cycles [83]. Reddy et al. utilized Moringa oleifera bark as a low-cost biosorbent for the biosorption of Ni(II) from aqueous phase [90]. The feasibility of developed biosorbent in field applications was studied by regenerating the biosorbent using 0.2 M HCl. The results demonstrated 98.02% recovery of used up adsorbent [90]. Khan and Rao carried out desorption studies by column process in a single metal ion system [92]. After adsorption experiment, the exhausted column was washed

Table 1 Various sorbents for the removal various toxic pollutants in aqueous systems

Biosorbent	Target pollutant	Adsorption efficiency (mg/g)	References
Red rose waste	Pb(II)	99.72	[70]
Rose petals waste	Pb(II), Cu(II)	119.92, 124.21	[71]
Orange peel	Pb(II), Zn(II)	209.8, 56.18	[72]
Cassava tuber bark waste	Zn(II)	83.30	[73]
Cedrusdeodara sawdust	Zn(II)	97.39	[74]
Cashew nut shell	Zn(II), Cu(II), Ni(II), Cd(II)	455.7, 406.6, 456.30, 436	[75]
Powdered wheat straw	Cd(II)	39.22	[76]
Rice bran	Ni(II)	46.51	[77]
Litchi chinensis seeds	Ni(II)	66.62	[78]
Cocoa pod husk	Ni(II)	20.10	[79]
Pine cone powder	Pb(II), Cu(II)	32.26, 26.32	[80]
Pineapple peel fiber	Pb(II), Cd(II),	70.29, 34.18	[81]
Cortex orange waste	Pb(II), Cu(II)	76.80, 67.20	[82]
Watermelon rind	Pb(II)	98.06	[83]
Onion skins	Cd(II)	200.00	[84]
Coffee grounds	Cd(II)	15.65	[85]
Water lily flower, Mangrove leaves,	Cr(VI)	8.44, 8.87	[86]
Cupressus lusitanica bark	Cr(VI)	87.5	[87]
Garden grass	Cu(II)	58.34	[88]
Palm oil fruit shell	Ni(II)	60.00	[89]
Moringa oleifera bark	Ni(II)	30.38	[90]
Caesalpinia bonducella seed	Ni(II)	188.67	[91]
Cucurbita moschata	Cu(II), Ni(II)	12.5, 15	[92]
Mannich base biosorbent	Pb(II)	60.5	[93]
Watermelon shell	Cu(II)	111.1	[94]
Organosolv lignin	Methylene blue	40.02	[95]
Lentinus edodes	Pb(II)	21.5	[96]
Citrus maxima peel, passion fruit shell, sugarcane bagasse	Pb(II), Cd(II), Cu(II), Ni(II)	169, 132, 84.0, 60.7 98.4, 48.6, 40.8, 25.8, 49.8, 26.3, 23.0, 16.1	[97]

(continued)

Table 1 (continued)

Biosorbent	Target pollutant	Adsorption efficiency (mg/g)	References
Modified Punica granatum L. peels	Pb(II)	371.36	[98]
Solanum melongena leaf powder	Pb(II)	71.42	[99]
Pomegranate peel	Cu(II)	30.12	[100]
Saccharum officinarum bagasse	As(III), As(V)	28.57, 34.48	[101]
Spent-grain	Cu(II)	10.47	[102]
Rice husk	As(V), Au, Cr(IV), Pb(II), Fe(III), Mn(II), Zn(II), Cd(II) malachite green and acid yellow	–	[103]
Zirconium-loaded okara'	Phosphates	44.13	[104]
Stevia leaves	Malachite green	284.45	[105]
Vigna radiata waste	Uranium	230	[106]
Saccharomyces cerevisiae loaded nano fibrous mats	Pb(II)	238.0	[107]

several times with double distilled water for the removal of traces of unadsorbed Cu(II) ions. After that, the column having adsorbed Cu(II) ions was treated with 0.1 M HCl acid to desorb metal ions [92]. Simultaneous uptake of As(III) and As(V) was demonstrated by Gupta et al. using green low-cost functionalized-biosorbent–Saccharum officinarum bagasse [101]. Regeneration and reusability studies showed up to 5 cycles utilization of developed adsorbent without much loss in sorption capacity [101]. Mullick and Neogi developed a potential biosorbent from used stevia leaves and utilized it for removal of malachite green from aqueous solution [105]. After one cycle of utilization, the spent biosorbents were air dried washed with distilled water, 0.1 M HNO_3 and 0.1 M NaOH solutions during the desorption process. It was found that the regenerated adsorbents had high adsorption efficiency thereby demonstrating the high potential of developed low-cost biosorbents [105].

9 Conclusion

The present chapter aims at the current progress associated with the desalination and heavy metal ion detoxification of water and wastewater using sorbents as adsorbents. These biosorbents were utilized for the removal of metal ions, organic compounds, and dyes from water samples. The adsorption capacities demonstrated in the various publications present effectiveness of the biosorbent for the specific type of metal species which in turn depends on various experimental parameters. The utilization of these sorbents for water and wastewater desalination and heavy metal detoxification presents excellent adsorption capacity towards metal ions, economic viability, non-hazardous nature, and biocompatibility. We imagine that biosorbents will get to be distinctly basic parts of open water treatment and modern frameworks.

10 Future Scenario

Even though wide studies in the literature on the uses of biosorbents for preconcentration, desalination, and heavy metal ion detoxification of water and wastewater have been done, still there are several research gaps that require being filled. Some of the essential characteristics that required to be detailed are computed as:

(1) The main focus is to choose an appropriate range of biosorbents to accomplish the maximum adsorption of a contaminant according to the adsorbent–adsorbate interactions.
(2) The optimization of various parameters of biosorbents having high functional groups on its surface is required to develop the removal efficiency towards a range of toxicants.
(3) The sorbents should be encouraged having low cost and high adsorption capability.
(4) The treatment potential of multi-component contaminants using these sorbent products is required for the large-scale uses of the adsorbents.
(5) More experiments should be performed on the effect of various co-contaminants due to the presence of phenols, dyes on adsorption of metal ions.

Therefore, development of the biosorbents having all the above-stated properties may present noteworthy benefits than presently developed commercially costly activated sorbents.

Acknowledgements We gratefully acknowledge support from the Ministry of Human Resource Development, Department of Higher Education, Government of India under the scheme of Establishment of Centre of Excellence for Training and Research in Frontier Areas of Science and Technology (FAST), for providing the necessary financial support to carry out this study vide letter No, F. No. 5–5/201 4–TS.Vll.

References

1. Pontius, F.W. 1990. *Water Quality and Treatment*. New York: McGraw-Hill Inc.
2. Faust, S.D., and O.M. Aly. 1987. *Adsorption Process for Water Treatment*. Stoneham: Butterworths Publishers.
3. Gupta, V.K., P.J.M. Carrott, M.M.L. Ribeiro Carrott, and Suhas. 2009. Low-cost adsorbents: Growing approach to wastewater treatment—A review. *Critical Reviews in Environmental Science and Technology* 39: 783–842.

4. Ahmaruzzaman, M., and V.K. Gupta. 2011. Rice husk and its ash as low-cost adsorbents in water and wastewater treatment. *Industrial and Engineering Chemistry Research* 50: 13589–13613.
5. Anastopoulos, I., and G.Z. Kyzas. 2014. Agricultural peels for dye adsorption: A review of recent literature. *Journal of Molecular Liquids* 200: 381–389.
6. Gautam, R.K. 2010. *Environmental Magnetism: Fundamentals and Applications*, 1st ed. Saarbrucken: LAP Lambert Academic Publishing.
7. Gautam, R.K., M.C. Chattopadhyaya, and S.K. Sharma 2013. Biosorption of Heavy Metals: Recent Trends and Challenges in Wastewater Reuse and Management, 305–322. Netherlands: Springer Netherlands.
8. O'Connell, D.W., C. Birkinshaw, and T.F. O'Dwyer. 2008. Heavy metal adsorbents prepared from the modification of cellulose: A review. *Bioresource Technology* 99: 6709–6724.
9. Zhou, Y.F., and R.J. Haynes. 2010. Sorption of heavy metals by inorganic and organic components of solid wastes: Significance to use of wastes as low-cost adsorbents and immobilizing agents. *Critical Reviews in Environment Science and Technology* 40: 909–977.
10. Babarinde, N.A., J.O. Babalola, A.O. Ogunfowokan, and A.C. Onabanjo. 2009. Kinetic, equi-librium, and thermodynamic studies of the biosorption of cadmium(II) from solution by Stereophyllum radiculosum. *Toxicological and Environmental Chemistry* 91: 911–922.
11. Fu, F., and Q. Wang. 2011. Removal of heavy metal ions from wastewaters: A review. *Journal of Environmental Management* 92: 407–418.
12. Rowbotham, A.L., L.S. Levy, and L.K. Shuker. 2000. Chromium in the environment: An evaluation of exposure of the UK general population and possible adverse health effects. *Journal of Toxicology and Environmental Health Part B: Critical Reviews* 3: 145–178.
13. Kimbrough, D.E., Y. Cohen, A.M. Winer, L. Creelman, and C. Mabuni. 1999. A critical assessment of chromium in the environment. *Critical Reviews in Environment Science and Technology* 29: 1–46.
14. Tewari, N., P. Vasudevan, and B.K. Guha. 2005. Study on biosorption of Cr(VI) by Mucor hiemalis. *Biochemical Engineering Journal* 23: 185–192.
15. Bhattacharyya, K.G., and S. Sen Gupta. 2006. Adsorption of chromium(VI) from water by clays. *Industrial and Engineering Chemistry Research* 45: 7232–7240.
16. Dayan, A.D., and A.J. Paine. 2001. Mechanisms of chromium toxicity, carcinogenicity and allergenicity: Review of the literature from 1985 to 2000. *Human & Experimental Toxicology* 20: 439–451.
17. Duruibe, J.O., M.O. Ogwuegbu, and J.N. Egwurugwu. 2007. Heavy metal pollution and human biotoxic effects. *International Journal of Physical Sciences* 2: 112–8.
18. Theophanides, T., and J. Anastassopoulou. 2002. Copper and carcinogenesis. *Crit. Rev. Oncol. Haematol.* 42: 57–64.
19. Kasprzak, K.S., F.W. Sunderman, and K. Salnikow. 2003. Nickel carcinogenesis. *Mutation Research* 533: 67–97.
20. Gupta, V.K., A. Rastogi, and A. Nayak. 2010. Biosorption of nickel onto treated alga (Oedo-gonium hatei): Application of isotherm and kinetic models. *Journal of Colloid and Interface Science* 342: 533–539.
21. Matlock, M.M., B.S. Howerton, and D.A. Atwood. 2002 Chemical precipitation of heavy metals from acid mine drainage. *Water Research* 36: 4757–4764.
22. Blöcher, C., J. Dorda, V. Mavrov, H. Chmiel, N.K. Lazaridis, and K.A. Matis. 2003. Hybrid flotation membrane filtration process for the removal of heavy metal ions from wastewater. *Water Research* 37: 4018–4026.
23. Rengaraj, S., C.K. Joo, Y. Kim, and J. Yi. 2003. Kinetics of removal of chromium from water and electronic process wastewater by ion exchange resins: 1200H, 1500H and IRN97H. *Journal of Hazardous Materials* 102: 257–27.

24. Hunsom, M., K. Pruksathorn, S. Damronglerd, H. Vergnes, and P. Duverneuil. 2005. Electro-chemical treatment of heavy metals (Cu(II), Cr(VI), Ni(II)) from industrial effluent and modelling of copper reduction. *Water Research* 39: 610–616.
25. Shaalan, H.F., M.H. Sorour, and S.R. Tewfik. 2001. Simulation and optimisation of a membrane system for chromium recovery from tanning wastes. *Desalination* 14: 315–324.
26. Rubio, J., M.L. Souza, and R.W. Smith. 2002. Overview of flotation as a wastewater treatment technique. *Minerals Engineering* 15: 139–155.
27. Liang, Y.J., L.Y. Chai, X.B. Min, C.J. Tang, H.J. Zhang, Y. Ke, and X.D. Xie. 2012. Hydro-thermal sulfidation and floatation treatment of heavy-metal-containing sludge for recovery and stabilisation. *Journal of Hazardous Materials* 217: 307–310.
28. Mohan, D., and K.P. Singh. 2002. Single- and multi-component adsorption of cadmium and zinc using activated carbon derived from bagasse–an agricultural waste. *Water Research* 36: 2304–2318.
29. Kobya, M., E. Demirbas, E. Senturk, and M. Ince. 2005. Adsorption of heavy metal ions from aqueous solutions by activated carbon prepared from apricot stone. *Bioresource Technology* 96: 1518–1521.
30. Chen, Y.H., and Y.D. Chen. 2011. Kinetic study of Cu(II) adsorption on nanosized $BaTiO_3$ and $SrTiO_3$ photocatalysts. *Journal of Hazardous Materials* 185: 168–173.
31. Wang, J., and C. Chen. 2006. Biosorption of heavy metals by Saccharomyces cerevisiae: A review. *Biotechnology Advances* 24: 427–451.
32. Song, J., H. Kong, and J. Jang. 2011. Adsorption of heavy metal ions from aqueous solution by poly rhodanine-encapsulated magnetic nanoparticles. *Journal of Colloid and Interface Science* 359: 505–511.
33. Volesky, B. 2007. Biosorption and me. *Water Research* 41: 4017–4029.
34. Das, N. 2010. Recovery of precious metals through biosorption—A review. *Hydromet-allergy* 103: 180–189.
35. Okoro, I.A., and S.O. Okoro. 2011. Agricultural byproducts as green chemistry absorbents for the removal and recovery of metal ions from wastewater environment. *Continental Journal of Water, Air and Soil Pollution* 2: 15–22.
36. Hong, G.B., and Y.K. Wang. 2017. Synthesis of low-cost adsorbent from rice bran for the removal of reactive dye based on the response surface methodology. *Applied Surface Science* 423: 800–809.
37. Liu, J., T. Mwamulima, Y. Wang, Y. Fang, S. Song, and C. Peng. 2017. Removal of Pb(II) and Cr(VI) from aqueous solutions using the fly ash-based adsorbent material-supported zero-valent iron. *Journal of Molecular Liquids*. doi:10.1016/j.molliq.2017.08.004.
38. Kılıç, M., C. Kırbıyık, Ö. Cepeliogullar, and A.E. Pütün. 2013. Adsorption of heavy metal ions from aqueous solutions by bio-char, a by-product of pyrolysis. *Applied Surface Science* 283: 856–862.
39. Daneshfozoun, S., M.A. Abdullah, and B. Abdullah. 2017. Preparation and characterization of magnetic biosorbent based on oil palm empty fruit bunch fibers, cellulose and Ceiba pentandra for heavy metal ions removal. *Industrial Crops and Products* 105: 93–103.
40. Wu, S.P., X.Z. Dai, J.R. Kan, F.D. Shilong, and M.Y. Zhu. 2016. Fabrication of carboxymethyl chitosan–hemicellulose resin for adsorptive removal of heavy metals from wastewater. Chinese Chemical Letters. http://dx.doi.org/10.1016/j.cclet.2016.11.015.
41. Malik, R., S. Dahiya, and S. Lata. 2017. *An experimental and quantum chemical study of removal of utmostly quantified heavy metals in wastewater using coconut husk: A novel approach to mechanism.* Int: International Journal of Biological Macromolecules. doi:10.1016/j.ijbiomac.2017.01.100.
42. Shakoor, S., and A. Nasar. 2017. Adsorptive treatment of hazardous methylene blue dye from artificially contaminated water using cucumis sativus peel waste as a low-cost adsorbent. *Groundwater for Sustainable Development*. doi:10.1016/j.gsd.2017.06.005.
43. Schiewer, S., and S.B. Patil. 2008. Pectin-rich fruit wastes as sorbents for heavy metal removal: Equilibrium and kinetics. *Bioresource Technology* 99: 1896–1903.

44. El-Said, A.G., N.A. Badawy, A.Y. Abdel-Aal, and S.E. Garamon. 2011. Optimisation parameters for adsorption and desorption of Zn(II) and Se(IV) using rice husk ash: Kinetics and equilibrium. *Ionics* 17: 263–270.
45. Ho, Y.S., and A.E. Ofomaja. 2006. Biosorption thermodynamics of cadmium on coconut copra meal as biosorbent. *Biochemical Engineering Journal* 30: 117–123.
46. Schiewer, S., and M. Iqbal. 2010. The role of pectin in Cd binding by orange peel biosorbents: A comparison of peels, depectinated peels and pectic acid. *Journal of Hazardous Materials* 177: 899–907.
47. Vaghetti, J.C., E.C. Lima, B. Royer, N.F. Cardoso, B. Martins, and T. Calvete. 2009. Pecan nutshell as biosorbent to remove toxic metals from aqueous solution. *Separation Science and Technology* 44: 615–644.
48. Alomá, I., M.A. Martín-Lara, I.L. Rodríguez, G. Blázquez, and M. Calero. 2012. Removal of nickel(II) ions from aqueous solutions by biosorption on sugarcane bagasse. *Journal of the Taiwan Institute of Chemical Engineers* 43: 275–281.
49. Liu, C., H.H. Ngo, W. Guo, and K.L. Tung. 2012. Optimal conditions for preparation of banana peels, sugarcane bagasse and watermelon rind in removing copper from water. *Bioresource Technology* 119: 349–354.
50. Pereira, F.V., L.V. Gurgel, and L.F. Gil. 2010. Removal of Zn(II) from aqueous single metal solutions and electroplating wastewater with wood sawdust and sugar cane bagasse modified with EDTA dianhydride (EDTAD). *Journal of Hazardous Materials* 176: 856–863.
51. Martín-Lara, M.Á., I.L. Rico, I.D. Vicente, G.B. García, and M.C. de Hoces. 2010. Modification of the sorptive characteristics of sugarcane bagasse for removing lead from aqueous solutions. *Desalination* 256: 58–63.
52. Pehlivan, E., H.T. Tran, W.K. Ouédraogo, C. Schmidt, D. Zachmann, and M. Bahadir. 2013. Sugarcane bagasse treated with hydrous ferric oxide as a potential adsorbent for the removal of As(V) from aqueous solutions. *Food Chemistry* 138: 133–138.
53. Krishnani, K.K., X. Meng, C. Christodoulatos, and V.M. Boddu. 2008. Biosorption mechanism of nine different heavy metals onto biomatrix from rice husk. *Journal of Hazardous Materials* 153: 1222–1234.
54. Khoramzadeh, E., B. Nasernejad, and R. Halladj. 2013. Mercury biosorption from aqueous solutions by Sugarcane Bagasse. *Journal of the Taiwan Institute of Chemical Engineers* 44: 266–269.
55. Pehlivan, E., T. Altun, and Ş. Parlayici. 2012. Modified barley straw as a potential biosorbent for removal of copper ions from aqueous solution. *Food Chemistry* 135: 2229–2234.
56. Miretzky, P., and A.F. Cirelli. 2010. Cr(VI) and Cr(III) removal from aqueous solution by raw and modified lignocellulosic materials: A review. *Journal of Hazardous Materials* 180: 1–19.
57. Muhamad, H., H. Doan, and A. Lohi. 2010. Batch and continuous fixed-bed column biosorption of Cd(II) and Cu(II). *Chemical Engineering Journal* 158: 369–377.
58. Asadi, F., H. Shariatmadari, and N. Mirghaffari. 2008. Modification of rice hull and sawdust sorptive characteristics for removing heavy metals from synthetic solutions and wastewater. *Journal of Hazardous Materials* 154: 451–458.
59. Bulut, Y. 2007. Removal of heavy metals from aqueous solution by sawdust adsorption. *Journal of Environmental Sciences* 19: 160–166.
60. Prado, A.G., A.O. Moura, M.S. Holanda, T.O. Carvalho, R.D. Andrade, I.C. Pescara, A.H. De Oliveira, E.Y. Okino, T.C. Pastore, D.J. Silva, and L.F. Zara. 2010. Thermodynamic aspects of the Pb adsorption using Brazilian sawdust samples: Removal of metal ions from battery industry wastewater. *Chemical Engineering Journal* 160: 549–555.
61. Palumbo, A.J., C.J. Daughney, A.H. Slade, and C.N. Glover. 2013. Impudence of pH and natural organic matter on zinc biosorption in a model lignocellulosic biofuel biorefinery effluent. *Bioresource Technology* 146: 169–175.
62. Šćiban, M., B. Radetić, Ž. Kevrešan, and M. Klašnja. 2007. Adsorption of heavy metals from electroplating wastewater by wood sawdust. *Bioresource Technology* 98: 402–409.

63. Saha, P.D., S. Chakraborty, and S. Chowdhury. 2012. Batch and continuous (fixed-bed column) biosorption of crystal violet by Artocarpus heterophyllus (jackfruit) leaf powder. *Colloids and Surfaces B* 92: 262–270.
64. Vijayaraghavan, K., and R. Balasubramanian. 2015. Is biosorption suitable for decontamination of metal-bearing wastewaters? A critical review on the state-of-the-art of biosorption processes and future directions. Journal of Environmental Management. http://dx.doi.org/10.1016/j.jenvman.2015.06.030.
65. Vijayaraghavan, K., and Y.S. Yun. 2008. Bacterial biosorbents and biosorption. *Biotechnology Advances* 26: 266–291.
66. Abdolali, A., W.S. Guo, H.H. Ngo, S.S. Chen, N.C. Nguyen, and K.L. Tung. 2014. Typical lignocellulosic wastes and by-products for biosorption process in water and wastewater treatment: A critical review. *Bioresource Technology* 160: 57–66.
67. Crini, G. 2005. Recent developments in polysaccharide-based materials used as adsorbents in wastewater treatment. *Progress in Polymer Science* 30: 38–70.
68. He, J., and J.P. Chen. 2014. A comprehensive review on biosorption of heavy metals by algal biomass: Materials, performances, chemistry, and modeling simulation tools. *Bioresource Technology* 160: 67–78.
69. Kumar, S.K., H.U. Dahmas, E.J. Won, J.S. Lee, and K.H. Shin. 2015. Microalgae—A promising tool for heavy metal remediation. *Ecotoxicology and Environmental Safety* 113: 329–352.
70. Haq Nawaz, B., K. Rubina, and H. Muhammad Asif. 2011. Biosorption of Pb(II) and Co(II) on red rose waste biomass. *Iranian Journal of Chemistry and Chemical Engineering (IJCCE)* 30: 80–88.
71. Manzoor, Q., R. Nadeem, M. Iqbal, R. Saeed, and T.M. Ansari. 2013. Organic acids pretreatment effect on Rosa bourbonia photo biomass for removal of Pb(II) and Cu(II) from aqueous media. *Bioresource Technology* 132: 446–452.
72. Feng, N., and X. Guo. 2012. Characterization of adsorptive capacity and mechanisms on adsorption of copper, lead and zinc by modified orange peel. *Transactions of the Nonferrous Metals Society of China* 22: 1224–1231.
73. Horsfall, J.M., A.A. Abia, and A.I. Spiff. 2006. Kinetic studies on the adsorption of Cd(II), Cu(II) and Zn(II) ions from aqueous solutions by Cassava (Munihot sculenta Cranz) tuber bark waste. *Bioresource Technology* 97: 283–291.
74. Mishra, V., C. Balomajumder, and V.K. Agarwal. 2012. Kinetics, mechanistic and thermodynamics of Zn(II) ion sorption: A modelling approach. *Clean—Soil Air Water* 40: 718–727.
75. Senthil Kumar, P., S. Ramalingam, R.V. Abhinaya, S.D. Kirupha, A. Murugesan, and S. Sivanesan. 2012. Adsorption of metal ions onto the chemically modified agricultural waste. *Clean – Soil Air Water* 40: 188–197.
76. Farooq, U., M.A. Khan, M. Athar, and J.A. Kozinski. 2011. Effect of modification of environmentally friendly biosorbent wheat (Triticum aestivum) on the biosorptive removal of cadmium(II) ions from aqueous solution. *Chemical Engineering Journal* 171: 400–410.
77. Zafar, M.N., R. Nadeem, and M.A. Hanif. 2007. Biosorption of nickel from protonated rice bran. *Journal of Hazardous Materials* 143: 478–485.
78. Flores-Garnica, J.G., L. Morales-Barrera, G. Pineda-Camacho, and E. Cristiani-Urbina. 2013. Biosorption of Ni(II) from aqueous solutions by Litchi chinensis seeds. *Bioresource Technology* 136: 635–643.
79. Njoku, V.O., A.A. Ayuk, E.E. Ejike, E.E. Oguzie, C.E. Duru, and O.S. Bello. 2011. Cocoapod husk as a low-cost biosorbent for the removal of Pb(II) and Cu(II) from aqueous solutions. *Australian Journal of Basic and Applied Sciences* 5: 101–110.
80. Ofomaja, A.E., E.B. Naidoo, and S.J. Modise. 2010. Biosorption of Cu(II) and Pb(II) onto potassium hydroxide treated pine cone powder. *Journal of Environmental Management* 91: 1674–1685.

81. Hu, X., M. Zhao, G. Song, and H. Huang. 2011. Modification of pineapple peel fibre with succinic anhydride for Cu(II), Cd(II) and Pb(II) removal from aqueous solutions. *Environmental Technology* 32: 739–746.
82. Kelly-Vargas, K., M. Cerro-Lopez, S. Reyna-Tellez, E.R. Bandala, and J.L. Sanchez-Salas. 2012. Biosorption of heavy metals in polluted water, using different waste fruit cortex. *Physics and Chemistry of the Earth* 37–39: 26–29.
83. Liu, C., H.H. Ngo, and W. Guo. 2012. Watermelon rind: Agro-waste or superior biosorbent. *Applied Biochemistry and Biotechnology* 167: 1699–1715.
84. Saka, C., Ö. Şahin, H. Demir, and M. Kahyaoğlu. 2011. Removal of lead from aqueous solutions using pre boiled and formaldehyde treated onion skins as a new adsorbent. *Separation Science and Technology* 46: 507–517.
85. Azouaou, N., Z. Sadaoui, A. Djaafri, and H. Mokaddem. 2010. Adsorption of cadmium from aqueous solution onto untreated coffee grounds: Equilibrium, kinetics and thermodynamics. *Journal of Hazardous Materials* 184: 126–134.
86. Elangovan, R., L. Philip, and K. Chandraraj. 2008. Biosorption of chromium species by aquatic weeds: Kinetics and mechanism studies. *Journal of Hazardous Materials* 152: 100–112.
87. Netzahuatl-Muñoz, A.R., Guillén-Jiménez F. de María, B. Chávez-Gómez, T.L. Villegas-Garrido, and E. Cristiani-Urbina. 2012. Kinetic study of the effect of pH on hexavalent and trivalent chromium removal from aqueous solution by Cupressus lusitanica bark. *Water, Air, and Soil Pollution* 223: 625–641.
88. Hossain, M.A., H.H. Ngo. W.S. Guo, and T. Setiadi. 2012. Adsorption and desorption of copper(II) ions onto garden grass. *Bioresource Technology* 121: 386–395.
89. Hossain, M.A., H.H. Ngo, W.S. Guo, and T.V. Nguyen. 2012. Palm oil fruit shells as biosorbent for copper removal from water and wastewater: Experiments and sorption models. *Bioresource Technology* 113: 97–101.
90. Reddy, D.H., D.K. Ramana, K. Seshaiah, and A.V. Reddy. 2011. Biosorption of Ni(II) from the aqueous phase by Moringa oleifera bark, a low-cost biosorbent. *Desalination* 268: 150–157.
91. Gutha, Y., V.S. Munagapati, S.R. Alla, and K. Abburi. 2011. Biosorptive removal of Ni(II) from aqueous solution by Caesalpinia bonducella seed powder. *Separation Science and Technology* 46: 2291–2297.
92. Khan, U., and R.A. Rao. 2017. A high activity adsorbent of chemically modified *Cucurbita moschata* (a novel adsorbent) for the removal of Cu(II) and Ni(II) from aqueous solution: Synthesis, characterization and metal removal efficiency. *Process Safety and Environment Protection* 107: 238–58.
93. Ge, Y., Q. Song, and Z.A. Li. 2015. A Mannich base biosorbent derived from alkaline lignin for lead removal from aqueous solution. *Journal of Industrial and Engineering Chemistry* 23: 228–234.
94. Banerjee, K., S.T. Ramesh, R. Gandhimathi, P.V. Nidheesh, and K.S. Bharathi. 2012. A novel agricultural waste adsorbent, watermelon shell for the removal of copper from aqueous solutions. *Iranica Journal of Energy & Environment* 3: 143–156.
95. Zhang, S., Z. Wang, Y. Zhang, H. Pan, and L. Tao. 2016. Adsorption of methylene blue on organosolv lignin from rice straw. *Procedia Environmental Sciences* 31: 3–11.
96. Chen, G.Q., G.M. Zeng, X. Tu, C.G. Niu, G.H. Huang, and W. Jiang. 2006. Application of a by-product of Lentinus edodes to the bioremediation of chromate contaminated water. *Journal of Hazardous Materials* 135: 249–255.
97. Chao, H.P., C.C. Chang, and A. Nieva. 2014. Biosorption of heavy metals on Citrus maxima peel, passion fruit shell, and sugarcane bagasse in a fixed-bed column. *Journal of Industrial and Engineering Chemistry* 20: 3408–3414.
98. Ay, Ç., A.S. Özcan, Y. Erdoğan, and A. Özcan. 2017. Characterization and lead (II) ions removal of modified Punica granatum L. peels. *International Journal of Phytoremediation* 19: 327–339.

99. Yuvaraja, G., N. Krishnaiah, M.V. Subbaiah, and A. Krishnaiah. 2014. Biosorption of Pb (II) from aqueous solution by Solanum melongena leaf powder as a low-cost biosorbent prepared from agricultural waste. *Colloids and Surfaces B: Biointerfaces* 114: 75–81.
100. Ben-Ali, S., I. Jaouali, S. Souissi-Najar, and A. Ouederni. 2017. Characterization and adsorption capacity of raw pomegranate peel biosorbent for copper removal. *Journal of Cleaner Production* 142: 3809–3821.
101. Gupta, A., S.R. Vidyarthi, and N. Sankararamakrishnan. 2015. Concurrent removal of As (III) and As (V) using green low cost functionalized biosorbent–Saccharum officinarum bagasse. *Journal of Environmental Chemical Engineering* 3: 113–121.
102. Lu, S., and S.W. Gibb. 2008. Copper removal from wastewater using spent-grain as biosorbent. *Bioresource Technology* 99: 1509–1517.
103. Chuah, T.G., A. Jumasiah, I. Azni, S. Katayon, and S.T. Choong. 2005. Rice husk as a potentially low-cost biosorbent for heavy metal and dye removal: An overview. *Desalination* 175: 305–316.
104. Nguyen, T.A., H.H. Ngo, W.S. Guo, J.L. Zhou, J. Wang, H. Liang, and G. Li. 2014. Phosphorus elimination from aqueous solution using 'zirconium loaded okara' as a biosorbent. *Bioresource Technology* 170: 30–37.
105. Mullick, A., and S. Neogi. 2016. Synthesis of potential biosorbent from used stevia leaves and its application for malachite green removal from aqueous solution: Kinetics, isotherm and regeneration studies. *RSC Advances* 6: 65960–65975.
106. Naeem, H., H.N. Bhatti, S. Sadaf, and M. Iqbal. 2017. Uranium remediation using modified Vigna radiata waste biomass. *Applied Radiation and Isotopes* 123: 94–101.
107. Xin, S., Z. Zeng, X. Zhou, W. Luo, X. Shi, Q. Wang, and Y. Du. 2017. Recyclable Saccharomyces cerevisiae loaded nanofibrous mats with sandwich structure constructing via bio-electrospraying for heavy metal removal. *Journal of Hazardous Materials* 324: 365–372.

Carbon Nanoadsorbents for Removal of Organic Contaminants from Water

Fernando Machado Machado and Éder Cláudio Lima

Abstract The removal of organic contaminants is of great concern in water treatment. This chapter elucidates the adsorption wastewater treatment processes using carbon nanoadsorbents with adsorbents. It is discussed the characteristics that make such nanostructures extremely interesting for adsorption process. In addition, a discussion of the main kinetics and isotherms models used to obtain information on the mechanisms and dynamics of the process is carried, as well as how these models are used and interpreted. Additionally, this chapter compiles relevant current knowledge about the experimental and theoretical adsorption activities of carbon nanotubes and graphene family as nanoadsorbents for removal of organic environmental pollutants. The accumulated data indicate that carbon nanomaterials can be successfully used for treating organic pollutants wastewater.

Keywords Adsorption · Textural properties · Nanomaterials · Nonlinear equilibrium and kinetic adsorption models · Thermodynamic calculation of entropy and enthalpy changes

Abbreviations

AC	Activated carbon
BET	Brunauer–Emmett–Teller
CNA	Carbon nanoadsorbents
CNT	Carbon nanotube
DFT	Density Functional Theory
EC	Emerging contaminants
EDC	Endocrine Disrupting Compounds

F.M. Machado (✉)
Centro de Desenvolvimento Tecnológico, Universidade Federal de Pelotas, Pelotas, RS 96010610, Brazil
e-mail: fernando.machado@hotmail.com.br

É.C. Lima
Instituto de Química, Universidade Federal do Rio Grande do Sul, Porto Alegre, RS 91501970, Brazil
e-mail: profederlima@gmail.com

© Springer International Publishing AG 2018
S. Bhardwaj Mishra and A.K. Mishra (eds.), *Bio- and Nanosorbents from Natural Resources*, Springer Series on Polymer and Composite Materials, https://doi.org/10.1007/978-3-319-68708-7_2

FLG–Few	Layer Graphene
GNS	Graphene nanosheet
GO	Graphene oxide
GOS	Graphene oxide nanosheet
MB	Methylene Blue
MWCNT	Multi-walled carbon nanotubes
OC	Organic contaminants
Q_{max}	Maximum adsorption capacity
R	Correlation coefficient
R^2	Coefficient of determination
R^2_{adj}	Adjusted coefficient of determination
SD	Standard deviation (root of mean square error)
SWCNT	Single-walled carbon nanotubes
rGO	Reduced Graphene Oxide

1 Introduction

One of the main threats to the survival of humanity in the coming decades is the possible shortage of drinking water. The scarcity of this natural resource is imminent, mainly due to the development models adopted by society for agriculture, livestock, and industry, in which concern for the environment has, for the most part, been relegated to the background. The misinformation and neglect of the treatment, combined the improper disposal of organic contaminants (OC), especially emerging contaminants (EC) such as pharmaceuticals and personal care products, endocrine disruptors, by-products of human activity, disinfectants, pesticides, synthetic dyes among others can result in irreversible damage to the environment and consequently the humans [1, 2]. The EC are potentially toxic substances, of which the presence and the effects on the environment are unknown [2–4].

Because of the strict regulations by the agencies and governmental organizations over the releases of pollutants to the environment, numerous methods for the treatment of effluents have been established [1, 2]. When compared with other methods for effluent treatment, the adsorption technique is the most promising procedure for effluent treatment [5–11].

What is Adsorption Technique?

The solids are designed because of the mutual attraction of the different atoms within the bulk. This attraction occurs due to necessitate of individual atoms have to achieve the electronic stability. In the interior of a bulk, the attractive forces are balanced among the numerous atoms making up the lattice. Nevertheless, the atoms that are at the solid surface have a deficient number of chemical bonds; therefore, these atoms are subjected to unbalanced forces. Since of this unbalanced nature, any particle that lands on the surface may be attracted by the solid. This is the adsorption phenomena, which is the process of concentrating solute at the surface of a solid by this attraction. Such phenomena, that occurring at different interfaces, we can consider this process in the following systems: liquid–gas, liquid–liquid, solid–gas, solid–liquid, and solid–solid [6, 7]. For the purposes of this chapter, we will deliberate only the solid–liquid system.

The adsorption is fundamentally a surface phenomenon involving two components: compound on which adsorption occurs, named as the "adsorbent," and the compound that is attached to the solid surface, defined as the "adsorbate." It is worth pointing out that the penetration by the adsorbate atoms or molecules into the bulk solid phase is typically defined as "absorption"; the terms "sorption," "sorbent," "sorptive," and "sorbate," are also used to denote mutually the adsorption and absorption phenomena, when both happen concurrently or cannot be distinguished [6, 7].

In a general way, the adsorption can be classified as physical (physisorption) or as chemical (chemisorption). Physisorption occurs in systems, which include only relatively weak intermolecular forces, on the other hand in the chemisorption, which involves fundamentally the formation of a chemical bond between the surface of the adsorbent and the sorbate molecule [5–7]. For a more accurate classification, other features must be considered, such as the origin of interactions, the specificity, and the heat of adsorption [5, 6].

From the viewpoint of environmental applications, the adsorption process is extremely valuable. When compared to other advanced techniques of purification or treatment of water, the adsorption has shown to be a superior route. This process is an efficient, easy, and simple method for the removal of OC from wastewater [5–11]. Here, the pollutants are transferred from an aqueous phase to the adsorbents surface [8, 9]. After adsorption of the pollutant, the treated effluent can be regenerated and reused making the process more economical and environmentally friendly [8–10]. Therefore, the determining factor for an efficient adsorption process is the appropriate choice of an adsorbent. When the choice of the adsorbent must be taken into account adsorption capacity, mechanical strength and chemical inertness of the adsorbent, as well as its availability and reusability [8–10]. Some nanomaterials especially the carbon-based, named carbon nanoadsorbents (CNA), satisfy most of these requirements, which make them extremely attractive for applications in water and effluents treatment by adsorption technique.

Why Use the Carbon Nanoadsorbents in Water Treatment?

In the last decades, the carbon nanomaterials, such as carbon nanotubes (CNTs) and graphene family, have been extensively analyzed for adsorption applications owing to their small sizes, structural diversity, low density, and good chemical stability [9, 10]. The textural properties of these nanomaterials, for example, surface area, pore volume, and average pore diameter, have inspired researches into the application of the CNA as possible adsorbents for water treatment [5, 8, 11, 12]. In addition, the lattice of materials favors the interaction between the adsorbate–adsorbent through of electrostatic interaction, π–π bonds, hydrophobic effect, hydrogen bonds, and covalent interaction [13, 14].

Another important point is that the chemical nature of CNA allows that their surfaces be easily altered by the physical and chemical process, thus enabling the directed improvement of their properties [5, 8]. Furthermore, the CNA can also be functionalized with metal oxides or metal such as ZnO and TiO_2 [15], Ag [16], Al/Al_2O_3 [17], and Fe/Fe_3O_4 [18]. The incorporation of metal or oxides has been shown to improve the adsorption properties of carbon nanomaterials [15–18].

These characteristics make the carbon nanomaterials attractive materials for the development of highly efficient and sensitive adsorbents devices for the removal of OC from wastewaters. Such characteristics will be discussed in the following section.

2 Carbon Nanoadsorbents

The carbon is the most versatile element in the periodic table, due to the characteristic and the number of bonds it can form with much different elements or with other carbon atoms [19]. This element can bind in a π-bond and a σ-bond while forming a molecule; the final molecular architecture depends on the level of hybridization of the C orbitals that is: a sp^1 hybridized C atom can make two σ-bonds and two π-bonds, a sp^2 hybridized C atom forms three σ-bonds and one π-bond, and a sp^3 hybridized C forms four σ-bonds [19, 20]. The nature and number of the bonds determine the geometry and consequently the properties of carbon allotropic forms [19, 20]. In the solid phase, the carbon can exist in many allotropic forms [19]. In this chapter, we will dwell on only the allotropes nanoscale or nanomaterials, particularly graphene family [21] and carbon nanotubes [22–24].

2.1 Carbon Nanotubes

The CNTs, single-walled (SWCNTs) and multi-walled carbon nanotubes (MWCNTs), have been widely studied since their discovery in 1991 [22] and 1993 [23, 24], respectively. These nanostructures have unique electronic and transport, mechanical, catalytic, adsorption properties that making them interesting for a variety of applications. Until the moment, CNTs have been produced through various routes by top-down and bottom-up design, mainly including laser ablation, arc discharge, and chemical vapor deposition (CVD) [20]. Compared with the other two approaches, the latter is considered as the most promising for easily scaled-up to batch-scale production due to economy and simplicity [20].

These CNTs have been shown to be a promising solution for the removal of OC through adsorption process [5]. The adsorption efficiency of CNTs depends basically on their textural properties [10, 11, 14]. The main textural properties affecting the removal of organic chemicals are pore size (micro- and mesopore), associated with the pore volume and the specific surface area [10, 11, 14, 25]. Numerous characterization techniques have shown the mesoporous nature of MWCNT and the microporous nature of SWCNT, combined with high surface area (by Brunauer–Emmett–Teller—BET method), ranging from 110.0 to 785.0 $m^2\ g^{-1}$ for MWCNT and 320.0 to 1587.0 $m^2\ g^{-1}$ for SWCNT [5, 8, 10, 11, 14, 25]. In the case of pore size, CNT can form large aggregates in aqueous environments. These large aggregates present diverse pore sizes, which increases their internal surface areas on the hydrophobic interior of the aggregates, allowing greater adsorption of OC [5, 10, 11, 14].

For SWCNT, the diameter and the number of the nanotubes in the bundle will affect largely the textural properties values, since in bundle the adsorption sites are different to those expected for an individual SWCNT [25]. The adsorption on SWCNT bundle can happen at four different sites: on the outside of the bundles, in the grooves formed at the contact between adjacent tubes, the inside of the nanotubes (pores) with open ends, and the interstitial channels between the tubes in bundles (see Fig. 1a) [5, 26, 27].

In contrast, the MWCNT are not in the bundle form and their adsorption sites consist of pores aggregated in the inner MWCNT surface with open ends and on the outer walls (see Fig. 1b) [5, 28]. Also, the occurrence of defects should be considered as reactive sites for adsorption [28]. The aggregated pores play an important role in the adsorption of large biological contaminants, for example, viruses and bacteria [28].

The structural properties of CNT allow a strong interaction with OC through non-covalent forces as, for example, hydrophobic interactions, hydrogen bonding, π–π stacking, electrostatic interaction, and van der Waals forces [11, 14, 25].

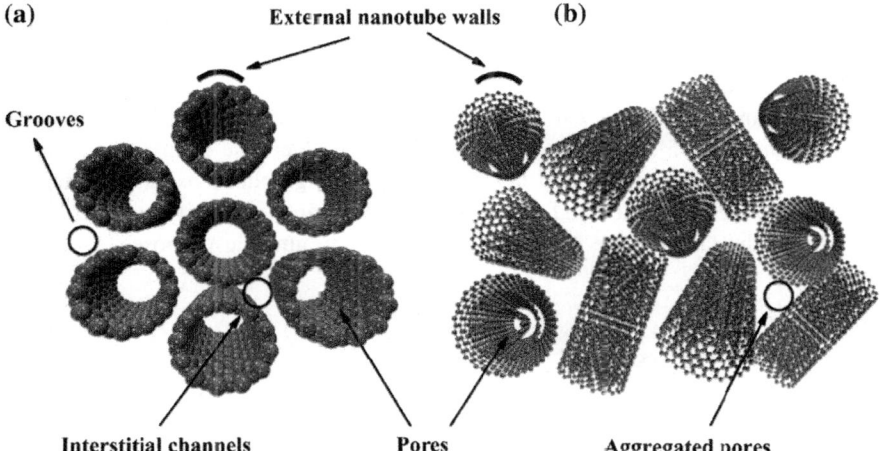

Fig. 1 Schematic model for the possible adsorption sites of **a** SWCNT bundle and for the **b** pores aggregated of MWCNT

Additionally, both nanotubes permit the incorporation on your walls of one or more functional groups, which may increase the selectivity of the resulting system [5, 29]. The presence of such functional groups also improves the solubility of the nanotubes in many fluid systems [5, 29].

2.2 Graphene Family

The graphene (Fig. 2a) is a two-dimensional sheet of sp^2 hybridized carbon, which was isolated experimentally for the first time by the group of researchers led by Andre Geim and Sir Konstantin Novoselov in 2004 [21]. Since its experimental obtaining, this new material with unique properties has fascinated the scientific community. Constituted by strength covalent bonds between carbon atoms, graphene has a very high tensile strength and hardness. In addition, this carbon allotrope has adjustable thermal and electrical conductivity, as well as excellent surface and optical feature through chemical marking that referring this nanomaterial to various practical applications [21, 30, 31]. Nowadays, the graphene family can be found either by top-down or by bottom-up methodology. The bottom-up method can produce defect-free graphene single sheets with excellent physical properties. Examples of this method are techniques such as epitaxial growth [30] and CVD [31]. Since the top-down methodology usually refers to the mechanical exfoliation of graphite to obtain few or single-sheet graphene. This was the route used by Geim and Novoselov [21]. Graphene can also be obtained in the oxidation of graphite by strong oxidizing agents followed by an exfoliation to give the graphene oxide (GO) (see Fig. 2b) [32–34]. This nanomaterial has been reductively processed by thermal, chemical, microwave, photothermal, photochemical, or microbial/bacterial to give the reduced graphene oxide (rGO) (see Fig. 2c) [35]. Numerous top-down methods are described in the literature for the synthesis of graphene, ranging from liquid-phase exfoliation [36] to electrochemical exfoliation [37] among others.

As noted above, depending on the route to obtain and characteristics of the produced nanomaterial, we can have many derivatives of graphene, to forming the graphene family [35]. The major types of "graphenes" can be defined: single-layer graphene, bilayer graphene, and few-layer graphene, graphene oxide, and reduced graphene oxide (for a detailed description of the graphene family see Ref. [35]). This being the last two the most exploited for environmental purposes.

Fig. 2 Schematic illustrating of **a** graphene, **b** GO and, **c** rGO

The GO is a monolayer material, chemically modified graphene synthetized by oxidation of the graphite and exfoliation that is accompanied by wide oxidative modification of the basal plane (see Fig. 2b) [5, 13, 38]. This has attracted considerable research interests due to its role as a precursor low cost for production of graphene-based materials, as well as its ease of production [5, 13, 38]. The GO has an extended layered structure with hydrophilic polar groups, for example, carbonyl (–C=O) and carboxyl groups (–COOH), situated at the edges of the sheet; hydroxyl (–OH), and epoxy groups (C–O–C), preferentially positioned in the basal plane [38–40]. Despite this acceptable configuration, the precise chemical structure of GO has been the subject of considerable debate, with uncertainty pertaining to both the type and distribution of oxygen-containing functional groups [38–40].

The GO has concerned substantial attention for applications such as adsorbent due to their textural properties, oxygen-containing functional groups, and high water solubility [5]. Experimentally, it is described that the surface area (by BET method) of as—synthetized GO ranges between 8.8 and 39.5 $m^2\ g^{-1}$ and the presence of large amount of mesopores [41–43]. Defects as the folding of sheets and wrinkling, in addition to the presence of curled sheets, contribute to the low surface area of GO [5].

The functional groups anchored on the GO surface, that provided the high negative charge density, can provide reactive sites for the adsorption of a diversity of adsorbates, particularly cationic species, such as synthetic dyes, mostly cationic dyes [39, 40, 44], and heavy metal ions [45–47]. This effect can be accentuated in higher pH solutions because in this condition the GO surface contains more negative charges, which pointedly increases its electrostatic interaction with positively charged of the cationic species [48]. Moreover, because of its aromatic structure, the GO can easily adsorb EC, for example, endocrine disrupting [49], and drugs [50, 51] on its surface due to interactions of the π–π stacking. It is noteworthy that the surface of GO can be decorated (or functionalized) with metal or metal oxides increasing a possible affinity for a specified adsorbate [16, 18].

As seen previously, the reduction of the GO results in the rGO. This is regularly considered as one kind of chemically derived graphene [13, 52], resulted of the incomplete deoxygenation of GO (see Fig. 2c) [35, 52]. There are different reduction processes of GO that result in different properties of rGO, which in turn affect the final performance of nanomaterials [5, 35].

About the textural properties, many authors highlight the high surface area as that obtained experimentally, which may vary from approximately 265.2 to 487.00 $m^2\ g^{-1}$ (obtained by BET method), and with high amounts of micro- and mesopores [13, 53, 54]. In addition, the rGO has a high wrinkling degree, numerous defects, and some residual oxygen-containing functional groups of the reduction process [13, 53, 54]. As well as the GO, the rGO can have your modified surface, increasing their affinity for a specific adsorbate [55, 56].

The characteristics described above make the rGO a promising CNA for remediation of containing drugs [13, 57], oil and organic solvent [58], endocrine interfering [59, 60], and synthetic dyes [39, 40]. The π–π stacking between sp^2 regions of rGO and aromatic structure of synthetic OC, as well as the electrostatic interactions between charged OC and the surface oxygen-containing groups of rGO, may also assist in the adsorption of these contaminants [13, 39, 40, 60, 61]. Unlike the GO, the rGO does not have as high negative surface charge, which makes this nanomaterial to be a fascinating adsorbent for anionic synthetic dyes [39, 40].

3 Kinetics and Equilibrium Models of Adsorption

Adsorption kinetic studies are important in treatment of aqueous effluents because they provide valuable information on the mechanism of the adsorption process. Many kinetic models were developed in order to find intrinsic kinetic adsorption constants. In this chapter, we will discuss the adsorption kinetic models based on chemical reactions (pseudo-first-order equation [62], pseudo-second-order equation [63], general order equation [64]), and the empiric models (Avrami fractional model [65], and Elovich-chemisorption model [66]).

At a fixed temperature, an adsorption isotherm describes the relationship between the sorption capacity of an adsorbent in relation to an adsorbate (q_e) and the adsorbate concentration remaining in solution after equilibrium is reached (C_e). The parameters from the adsorption equilibrium models provide useful information on the adsorption mechanism. There are several different isotherm models for describing the equilibrium of an adsorbate on an adsorbent. The most employed and discussed in the literature is the Langmuir model [67]. Other isotherm equations such as Freundlich isotherm [68], Sips isotherm [69], Liu isotherm [70], Redlich–Peterson isotherm [71] are also well discussed in the literature.

3.1 Kinetic Adsorption Models

3.1.1 Kinetic Models Based on Chemical Reactions

The study of adsorption kinetic is important in the treatment of aqueous effluents using nanomaterials because it provides relevant information on the adsorption steps and the mechanism of adsorption.

Many kinetic models were developed based on the kinetics of chemical reactions. The first kinetic model of adsorption is described using the equations developed by Lagergren [62]. A simple kinetic analysis of adsorption is the pseudo-first-order equation in the form of Eq. 1.

$$\frac{dq}{dt} = k_1 \cdot (q_e - q_t) \qquad (1)$$

where q_t is the amount of adsorbate adsorbed at time t (mg g^{-1}), q_e is the equilibrium adsorption capacity (mg g^{-1}), k_1 is the pseudo-first-order rate constant (min^{-1}), and t is the contact time (min; h). The integration of Eq. 1 with initial conditions, $q_t = 0$ at $t = 0$, and $q_t = q_t$ at $t = t$ leads to Eq. 2.

$$\ln(q_e - q_t) = \ln(q_e) - k_1 \cdot t \qquad (2)$$

A nonlinear rearrangement of Eq. 2 gives Eq. 3.

$$q_t = q_e \cdot [1 - \exp(-k_1 \cdot t)] \qquad (3)$$

Equation 3 is known as pseudo-first-order kinetic adsorption model.

In addition, a pseudo-second-order equation [63] based on equilibrium adsorption capacity is shown in Eq. 4.

$$\frac{dq_t}{dt} = k_2 \cdot (q_e - q_t)^2 \qquad (4)$$

where k_2 is the pseudo-second-order rate constant (g mg^{-1} min^{-1}). The integration of Eq. 4 with initial conditions, $q_t = 0$ at $t = 0$, and $q_t = q_t$ at $t = t$ leads to Eq. 5.

$$q_t = \frac{k_2 \cdot q_e^2 \cdot t}{1 + q_e \cdot k_2 \cdot t} \qquad (5)$$

Equation 5 is known as pseudo-second-order kinetic adsorption model.

The pseudo-first-order and pseudo-second-order are the most commonly employed kinetic models for describing adsorption process based on chemical reactions. Another approach to this theme is described below.

The exponents of rate laws of chemical reactions are usually independent of the stoichiometric coefficients of chemical equations but are sometimes related [72, 73]. This assertion implies that the order of a chemical reaction depends solely on the experimental data [72, 73]. Adsorption process, which is considered to be the rate-determining step helps in establishing the general rate law equation of adsorption process [25, 64, 72, 73]. Attention is now focused on the change in the effective number of active sites at the surface of adsorbent during adsorption instead of concentration of adsorbate in the bulk solution [72]. Applying reaction rate law to Eq. 6 gives adsorption rate expression [25, 64, 72]:

$$\frac{dq}{dt} = k_N \cdot (q_e - q_t)^n \qquad (6)$$

where k_N is the rate constant, q_e is the amount of adsorbate adsorbed by adsorbent at equilibrium, q_t is the amount of adsorbate adsorbed by adsorbent at a given time, t, and n is the order of adsorption with respect to the effective concentration of the adsorption active sites present on the surface of adsorbent [72]. Application of universal rate law to adsorption process led to Eq. 6, which can be used without assumptions. Theoretically, the exponent n in Eq. 6 can be an integer or non-integer rational number [25, 64, 72, 73].

Equation 7 describes the number of the active sites (θ_t) available on the surface of adsorbent for adsorption [25, 64, 72].

$$\theta_t = 1 - \frac{q_e}{q_t} \tag{7}$$

Equation 8 describes the relationship between the variable (θ_t) and rates of adsorption.

$$\frac{d\theta_t}{dt} = -k\theta_t^n \tag{8}$$

By definition [72]

$$k = k_N(q_e)^{n-1} \tag{9}$$

The value of θ_t will be equal to 1 if an adsorbent has not adsorbed [72]. The value of θ_t decreases during adsorption process. θ_t approaches a fixed value when adsorption process reaches equilibrium. $\theta_t = 0$ for a saturated adsorbent [72]. Equation 8 gives Eq. 10 after integration.

$$\int_1^\theta \frac{d\theta_t}{\theta_t^n} = -k \int_0^t dt \tag{10}$$

After integration of Eq. 10, it results in Eq. 11.

$$\frac{1}{1-n} \cdot \left[\theta_t^{1-n} - 1\right] = -kt \tag{11}$$

Making a mathematical rearrangement of Eq. 11, it gives Eq. 12.

$$\theta_t = \left[1 - k(1-n) \cdot t\right]^{1/1-n} \tag{12}$$

Substituting Eqs. 7 and 9 into Eq. 12, it yields Eq. 13.

$$q_t = q_e - \frac{q_e}{\left[k_N(q_e)^{n-1} \cdot t \cdot (n-1) + 1\right]^{1/1-n}} \tag{13}$$

Equation 13 is regarded as the general-order kinetic equation of adsorption process, which is valid for $n \neq 1$ [25, 64, 72].

A special case of Eq. 8 is the pseudo-first-order kinetic model ($n = 1$) [12, 14].

$$\frac{d\theta_t}{dt} = -k\theta_t^1 \quad (14)$$

Integration of Eq. 14 gives Eq. 15.

$$\theta_t = \exp(-k \cdot t) \quad (15)$$

Substitution of Eq. 7 into Eq. 15, and put $k = k_1$ gives pseudo-first-order kinetic model as shown in Eq. 16.

$$q_t = q_e[1 - \exp(-k_1 \cdot t)] \quad (16)$$

Pseudo-first-order kinetic equation is a special case of general kinetic model of adsorption. It must be noted that Eq. 16 is the same as Eq. 3 using the adsorption rate expression [72].

For the general-order kinetic model, when $n = 2$, the pseudo-second-order kinetic model is a special case of Eq. 13 [8].

$$q_t = q_e - \frac{q_e}{[k_2(q_e) \cdot t + 1]} \quad (17)$$

Making a mathematical rearrangement on Eq. 17 gives Eq. 28.

$$q_t = \frac{q_e^2 k_2 t}{[k_2(q_e) \cdot t + 1]} \quad (18)$$

Equation 18 is the pseudo-second-order kinetic adsorption model, which is exactly the same as Eq. 5. Therefore, the general-order adsorption Eq. 13 could give rise to pseudo-second-order when $n = 2$, while the pseudo-first-order is obtained from Eq. 8 (adsorption rate expression).

Figures 3, 4, and 5 present respectively the kinetic curves of adsorption of Reactive Violet 5 onto cocoa activated carbon [74], using pseudo-first-order, pseudo-second-order, and general-order kinetic models.

3.1.2 Empiric Models

The Elovich equation is generally applied to chemisorption kinetics [75]. The equation has been used satisfactorily for some chemisorption processes [72] and has been found to cover a wide range of slow adsorption rates. The same equation is often valid for systems in which the adsorbing surface is heterogeneous. The Elovich equation is given in Eq. 19.

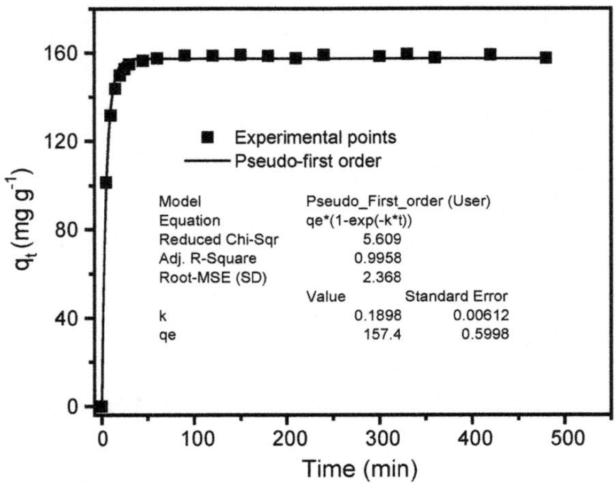

Fig. 3 Pseudo-first-order kinetic adsorption model of RV-5 dye using cocoa shell activated carbon. Initial pH, 2.0; temperature, 25 °C; initial concentration of the adsorbate, 400.0 mg L^{-1}

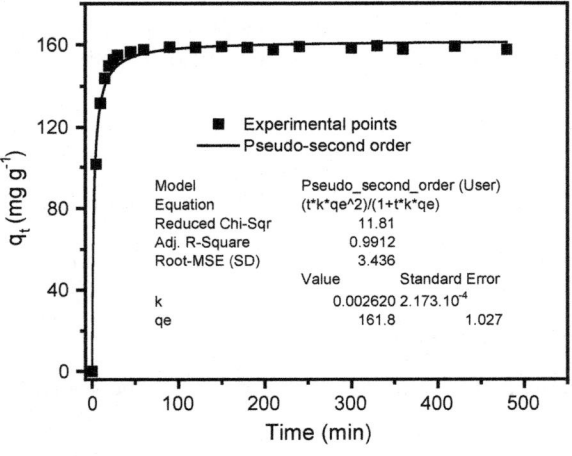

Fig. 4 Pseudo-second-order kinetic adsorption model of RV-5 dye using cocoa shell activated carbon. Initial pH, 2.0; temperature, 25 °C; initial concentration of the adsorbate, 400.0 mg L^{-1}

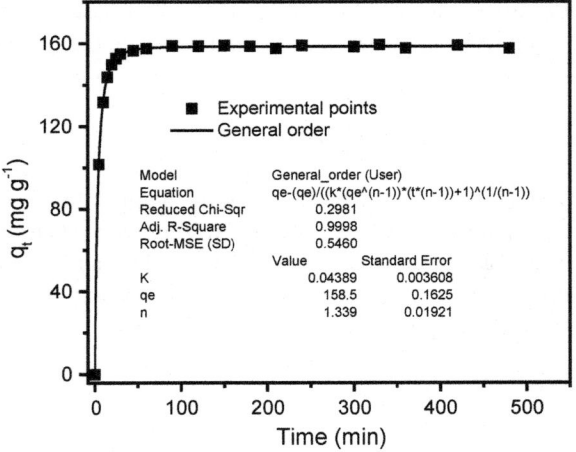

Fig. 5 General-order kinetic adsorption model of RV-5 dye using cocoa shell activated carbon. Initial pH, 2.0; temperature, 25 °C; initial concentration of the adsorbate, 400.0 mg L^{-1}

$$\frac{dq_t}{dt} = \alpha \exp(-\beta q_t) \qquad (19)$$

Integrating Eq. 19 using boundary conditions; $q_t = 0$ at $t = 0$ and $q_t = q_t$ at $t = t$ gives Eq. 20.

$$q_t = \frac{1}{\beta}\ln(t + t_o) - \frac{1}{\beta}\ln(t_o) \qquad (20)$$

where α is the initial adsorption rate (mg g^{-1} min^{-1}) and β is related to the extent of surface coverage and the activation energy involved in chemisorption (g mg^{-1}) and $t_o = 1/\alpha\beta$.

If t is much larger than t_o, the kinetic equation can be simplified as Eq. 21.

$$q_t = \frac{1}{\beta}\ln(\alpha \cdot \beta) + \frac{1}{\beta}\ln(t) \qquad (21)$$

Equation 21 is known as Elovich-chemisorption kinetic adsorption model.

In Fig. 6 is illustrated the Elovich kinetic model for adsorption of Cr(VI) ions onto Brazilian-pine-fruit shell [75].

The determination of some kinetic parameters, possible changes of the adsorption rates as function of the initial concentration and the adsorption time as well as the determination of fractional kinetic orders, still lacks in the kinetic adsorption models [65, 76]. In this way, an alternative Avrami kinetic equation to find a correlation between good experimental and calculated data was early proposed [65, 76]. The adsorption should now be visualized using Avrami's exponential function, which is an adaptation of kinetic thermal decomposition modeling [76]:

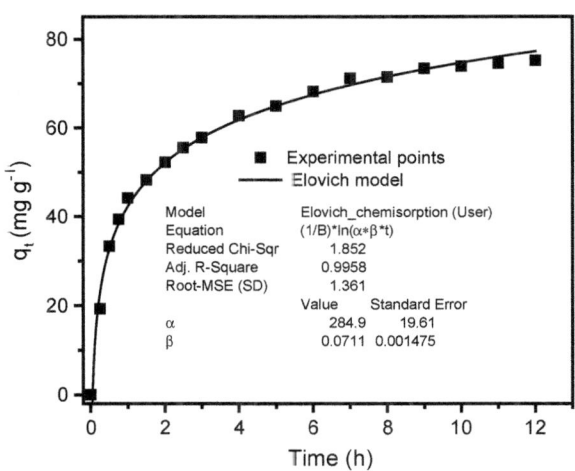

Fig. 6 Elovich kinetic adsorption model of Cr(VI) onto Brazilian-pine-fruit shell biosorbent. Initial pH, 2.0; temperature, 25 °C; initial concentration of the Cr(VI), 500.0 mg L^{-1}

$$\alpha = 1 - \exp[(-k_{AV} \cdot t)]^{n\,AV} \qquad (22)$$

where α is adsorption fraction (q_t/q_e) at time t, k_{AV} is the Avrami kinetic constant (min^{-1}), and n_{AV} is a fractional adsorption order, which is related to the adsorption mechanism [65, 76]. By inputting α in Eq. 22, the Avrami kinetic equation could be written as Eq. 23.

$$q_t = q_e \cdot \{1 - \exp[-(k_{AV} \cdot t)]^n\} \qquad (23)$$

In Fig. 7 is presented the Avrami fractional-order kinetic model of adsorption of 500 mg L^{-1} phenol onto wood sawdust activated carbon [77].

Usually for testing a given kinetic and equilibrium model, some statistical functions are tested. The best-fit model is the one with the lowest value of standard deviation (SD) and the one in which the value of adjusted coefficient of determination (R^2_{adj}) is closer to unity. Equations 24–27 depict the expressions of reduced chi-square, SD, coefficient of determination (R^2), and R^2_{adj}, respectively.

$$\text{Reduced Chi - squared} = \sum_i^n \frac{(q_{i,\exp} - q_{i,\text{model}})^2}{n_p - p} \qquad (24)$$

$$SD = \sqrt{\left(\frac{1}{n_p - p}\right) \cdot \sum_i^n (q_{i,\exp} - q_{i,\text{model}})^2} \qquad (25)$$

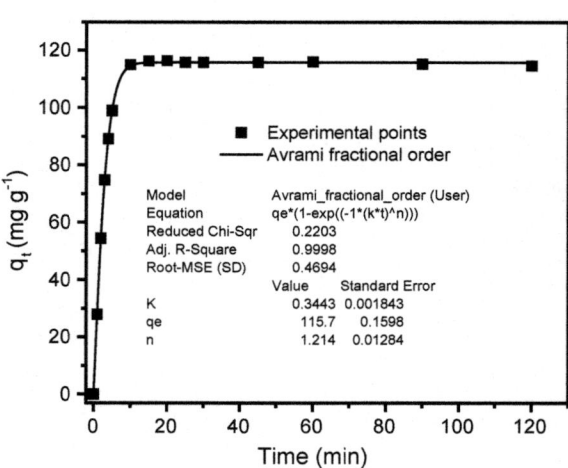

Fig. 7 Avrami fractional-order kinetic adsorption model of phenol onto wood sawdust activated carbon. Initial pH, 7.0; temperature, 25 °C; initial concentration of phenol, 500.0 mg L^{-1}

$$R^2 = \left[\frac{\sum_i^{np} (q_{i,\text{exp}} - \bar{q}_{i,\text{exp}})^2 - \sum_i^{np} (q_{i,\text{exp}} - q_{i,\text{model}})^2}{\sum_i^n (q_{i,\text{exp}} - \bar{q}_{i,\text{exp}})^2} \right] \tag{26}$$

$$R_{\text{adj}}^2 = 1 - (1 - R) \cdot \left(\frac{n_p - 1}{n_p - p - 1} \right) \tag{27}$$

where q_i model is each value of q predicted by the fitted model, $q_{i,\text{exp}}$ is each value of q measured experimentally, $\bar{q}_{\text{exp}}$ is the average of q experimentally measured, n_p is the number of experiments performed, and p is the number of parameters of the fitted model.

The reduced chi-squared is the residual sum of squares divided by the degree of freedom $(n_p - p)$ (Eq. 24). The SD is the square root of reduced chi-squared (Eq. 25). Both Eqs. 24 and 25 are very useful for evaluation of point-to-point of a given kinetic or equilibrium adsorption model; this is because for each experimental point there is a point in the model that corresponds exactly to the point on the curve (model). The lower the reduced chi-squared and the SD, the lower the difference between the values of experimental q and theoretical q; therefore, the best fit is expected. However, it should be taken into account that it is not possible to compare the values of reduced chi-squared and SD among several kinetic and equilibrium isotherms that present other different concentrations or other conditions since the values of SD and chi-square tend to increase as the concentration increases. On the contrary, for the same set of experimental data, the values of reduced chi-squared and SD are useful to ascertain the best model since R^2 and R_{adj}^2 are of low sensitivity (their values are limited to unity) [72].

The R^2 and R_{adj}^2 in Eqs. 26 and 27, respectively, are very useful parameters to evaluate kinetic and equilibrium adsorption fits. Limitedly, their values are between 0 and 1. Values of R^2 and R_{adj}^2 that are closer to 1 means that the model has a better fit. It is important to note that $\bar{q}_{i,\text{exp}}$ of Eq. 26 is the average of all experimental data (q). If the range of q values is too large, $\bar{q}_{i,\text{exp}}$ could distort the interpretation of the fit. If there are equidistant values of $q_{i,\text{exp}}$ from the average value, the values of R^2 tend to 1. Therefore, the analysis of a kinetic could not only be based on the values of R^2 [72]. In the same way, comparison of 2-parameter models (pseudo-first-order, pseudo-second-order, Elovich-chemisorption) with 3-parameter models (general order, Avrami fractional order) is not possible because the equations with higher number of parameters have tendency to exhibit R^2 values closer to 1. In these cases, it is recommended to use R_{adj}^2 [72]. This statistical parameter is used to penalize the models with more parameters in order to really know if the best fitting (R_{adj}^2) is due to the advantage of presenting more terms in the equation (mathematical advantage), or alternatively, the equation is physically closer to the reality of the system. R_{adj}^2 is, therefore, a very good parameter for evaluating a given kinetic and equilibrium of adsorption [72].

3.2 Equilibrium Isotherm Models

3.2.1 Langmuir Isotherm Model

The Langmuir [67] isotherm is based on the following assumptions:
- Adsorbates are adsorbed at a fixed number of well-defined sites;
- A monolayer of the adsorbate is formed over the surface of the adsorbent when it gets saturated;
- Each site can hold only one adsorbate specie;
- All sites are energetically equivalent;
- Interactions between the adsorbate species do not exist.

The Langmuir isotherm equation is depicted by Eq. 28.

$$q_e = \frac{Q_{\max} \cdot K_L \cdot C_e}{1 + K_L \cdot C_e} \tag{28}$$

where q_e is the amount of adsorbate adsorbed at the equilibrium (mg g^{-1}), C_e is the supernatant adsorbate concentration at the equilibrium (mg L^{-1}), K_L is the Langmuir equilibrium constant (L mg^{-1}), and $Q_{\max}$ is the maximum adsorption capacity of the adsorbent (mg g^{-1}) assuming a monolayer of adsorbate uptake by the adsorbent.

In Fig. 8 is presented the Langmuir isotherm for adsorption of sodium diclofenac onto rGO adsorbent [13].

The R^2_{adj} for nonlinear Langmuir isotherm was 0.9618, the $Q_{\max}$ was 69.41 mg g^{-1} and the Langmuir equilibrium constant was 0.1048 L mg^{-1}. The total standard deviation of the fitting was 3.750 mg g^{-1}.

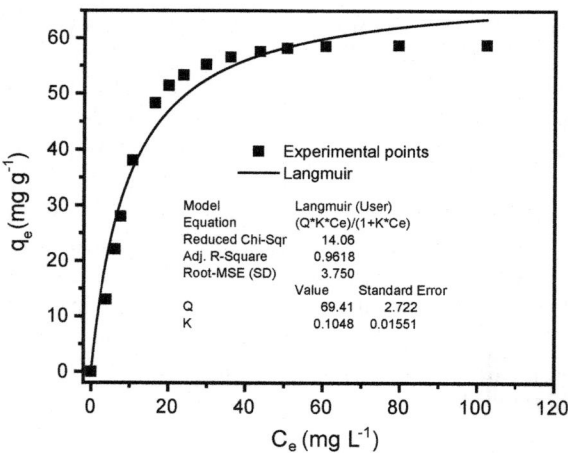

Fig. 8 Langmuir adsorption equilibrium isotherm of sodium diclofenac onto rGO at 25 °C and pH 10.0

3.2.2 Freundlich Isotherm Model

Freundlich [68] isotherm model is an exponential equation and assumes that the concentration of adsorbate on the adsorbent surface increases as the adsorbate concentration increases. Theoretically, using this expression, an infinite amount of adsorption will occur. Similarly, the model assumes that the adsorption could occur via multiple layers instead of a single layer. The equation has a wide application in heterogeneous systems. Equation 29 shows the Freundlich isotherm model;

$$q_e = K_F \cdot C_e^{1/n_F} \qquad (29)$$

where K_F is the Freundlich equilibrium constant (mg g^{-1}(mg L^{-1})$^{-1/n}$), and n_F is the Freundlich exponent (dimensionless).

Figure 9 shows the Freundlich isotherm plot for adsorption of sodium diclofenac onto rGO adsorbent [13].

The R^2_{adj} for nonlinear Freundlich isotherm was 0.8417 while the Freundlich equilibrium constant was 18.83 (mg g^{-1}(mg L^{-1})$^{-n/n_F}$). The total standard deviation of the fitting was 7.384 mg g^{-1}.

3.2.3 Sips Isotherm Model

Sips model, an empirical model, consists of the combination of the Langmuir and Freundlich isotherm models. The Sips [69] model takes the following form:

$$q_e = \frac{Q_{max} \cdot K_S \cdot C_e^{n_S}}{1 + K_S \cdot C_e^{n_S}} \quad \text{where } 0 < n_s \leq 1 \qquad (30)$$

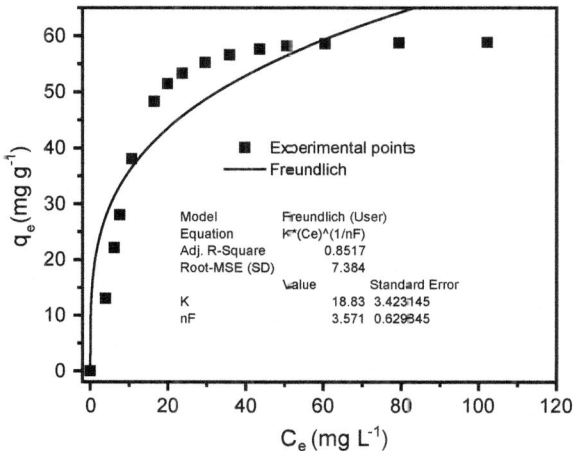

Fig. 9 Freundlich adsorption equilibrium isotherm of sodium diclofenac onto rGO at 25 °C and pH 10.0

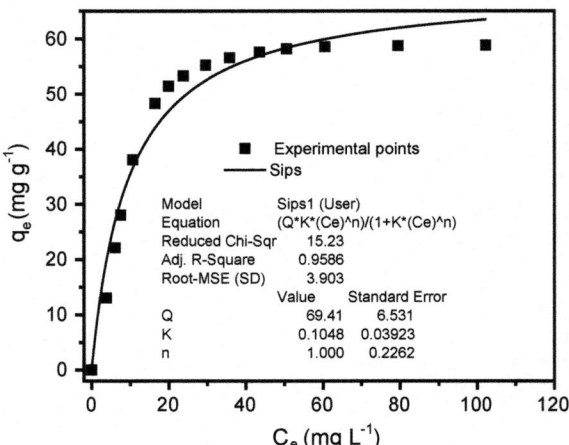

Fig. 10 Sips adsorption equilibrium isotherm of sodium diclofenac onto rGO at 25 °C and pH 10.0

In Eq. 30, K_S is the Sips equilibrium constant (mg L^{-1})$^{-1/n}$, Q_{max} is the Sips maximum adsorption capacity (mg g^{-1}), and n_S is the exponent. It is assumed that the $1/n_S$ should be ≤ 1 for integration purpose [69].

Although several works in the literature using Sips isotherm really do not take into account this consideration.

The Fig. 10 shows the Sips isotherm curve for adsorption of sodium diclofenac onto rGO adsorbent [13].

At low adsorbate concentrations, Sips equation relatively reduces to the Freundlich isotherm, but it predicts a monolayer adsorption capacity characteristic of the Langmuir isotherm at high adsorbate concentrations.

It was observed that the Sips exponent was restricted to 1 [69], in this manner, this isotherm model has the same parameters as the Langmuir isotherm; however, the R_{adj}^2 and the SD of the Sips isotherm were worse than those of the Langmuir isotherm. In that case, when $n_s = 1$, the Langmuir expression is preferred because the Sips isotherm has three parameters while Langmuir isotherm has just two. This parametric difference worsens the values of R_{adj}^2 and SD of the Sips isotherm.

3.2.4 Liu Isotherm Model

The Liu isotherm model [70] is a combination of the Langmuir and Freundlich isotherm models, but the monolayer assumption of Langmuir model and the infinite adsorption assumption that originates from the Freundlich model are discarded. The Liu model predicts that the active sites of the adsorbent cannot possess the same energy.

Therefore, the adsorbent may present active sites preferred by the adsorbate molecules for occupation [70], however, saturation of the active sites should occur unlike in the Freundlich isotherm model. Equation 31 defines the Liu isotherm model:

$$q_e = \frac{Q_{\max} \cdot (K_g \cdot C_e)^{n_L}}{1 + (K_g \cdot C_e)^{n_L}} \quad (31)$$

where K_g is the Liu equilibrium constant (L mg^{-1}); n_L is dimensionless exponent of the Liu equation, $Q_{\max}$ is the maximum adsorption capacity of the adsorbent (mg g^{-1}). Contrary to the Sips isotherm, n_L could assume any positive value.

Fig. 11 shows the Liu isotherm plot for adsorption of sodium diclofenac onto rGO adsorbent [13].

From the fit, the R^2_{adj} obtained was 0.9995, which is very good for a nonlinear isotherm. Similarly, the SD for Liu isotherm model was only 0.4121 mg g^{-1}. This value was 9.10, 17.9, 9.47 times lower than the SD values of Langmuir, Freundlich, and Sips isotherms models, respectively, indicating that this isotherm model was a better fit to the experimental equilibrium data [72]. The advantages of Liu isotherm model (a 3-parameter isotherm) over the Sips isotherm model is that the exponent of Liu isotherm could admit any positive value unlike the exponent of Sips that is limited to $0 \leq n_S \leq 1$.

3.2.5 Redlich–Peterson Isotherm Model

This is an empirical equation that describes an equilibrium isotherm as shown in Eq. 32 [71].

$$q_e = \frac{K_{RP} \cdot C_e}{1 + a_{RP} \cdot C_e^g} \quad \text{where } 0 < g \leq 1 \quad (32)$$

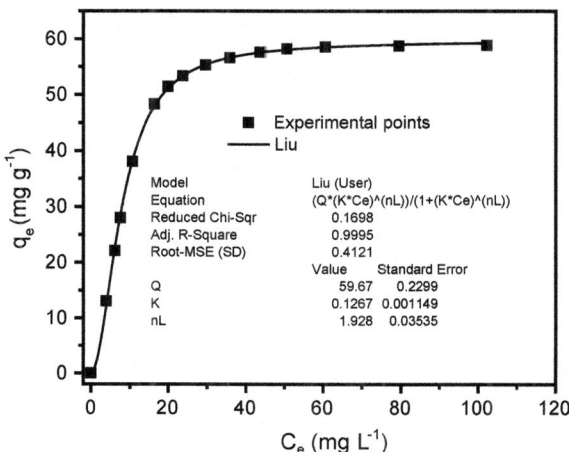

Fig. 11 Liu adsorption equilibrium isotherm of sodium diclofenac onto rGO at 25 °C and pH 10.0

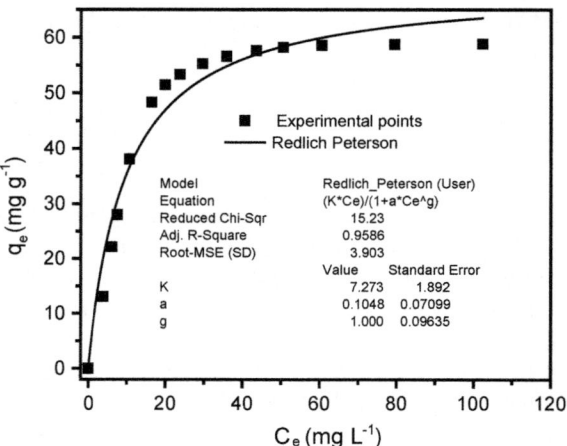

Fig. 12 Redlich–Peterson adsorption equilibrium isotherm of sodium diclofenac onto rGO at 25 °C and pH 10.0

where K_{RP} and a_{RP} are Redlich–Peterson constants with the respective units of L g^{-1} and (mg L^{-1})$^{-g}$, and g is the Redlich–Peterson exponent (dimensionless) whose value should be ≤ 1. This equation becomes linear at a low surface coverage ($g = 0$) and reduces to a Langmuir isotherm when $g = 1$. Fig. 12 presents the Redlich–Peterson isotherm curve for adsorption of sodium diclofenac onto rGO adsorbent [13].

4 Removal of Organic Contaminants by Carbon Nanoadsorbents

As highlighted throughout this chapter, the carbon nanostructures can play an important role in the adsorption process, because of their good chemical stability, textural properties, low density, and suitability for large-scale production. In this section, we present an overview of the current and available research on carbon-based adsorbents, for example, graphene family and carbon nanotubes for removal of organic contaminants such as synthetic dyes, as well as pharmaceuticals and endocrine disruptors from wastewater.

4.1 Adsorption of Dyes

Numerous industries such as textile and leather, feedstuffs, paper, drugs, and cosmetics, among others apply dyes and pigments to color their products. The dyeing process involves several steps which generate large volumes of colored effluents [11, 14, 25]. Throughout manufacturing processes, about of 10–60% of dyes are lost, generating large quantities of colored wastewaters [11, 14, 25]. The presence of pigments and dyes in water is highly undesirable and represents a worrisome environmental problem [11, 14, 25].

Recently, several studies have explored the application potential of CNA for adsorptive decolorization of dye wastewaters. In this context, some batch adsorption studies have been directed to evaluate the dye adsorption behavior on CNT.

SWCNTs and MWCNTs were used as adsorbents for the removal of Alizarin Red S dye from aqueous solutions [14]. Both CNT showed a fast adsorption kinetics, and the contact time was fixed at 65 and 100 min, respectively, for SWCNT and MWCNT. The general-order kinetic model showed to best fit the adsorption data. A good fitting of adsorption isotherms was obtained using a Liu model. The maximum sorption capacities of Alizarin Red S dye onto SWCNTs and MWCNTs were 312.5 and 135.2 mg g^{-1}, respectively, at 318 K (for pH 2). Through the thermodynamic calculations, the authors demonstrated that the adsorption process was exothermic and spontaneous for all temperatures. In addition, through of change in enthalpy, the authors determined the interaction of the dye with both CNTs was dominated by electrostatic attraction. Theoretical calculations were used to study the interaction of dye with (5, 5) and (8, 0) SWCNTs with and without vacancy, and (16, 0) and (25, 0) SWCNTs without vacancy. The results from ab initio calculations designated that electrostatic interaction may be responsible for the adsorption of ARS by SWCNT. These results agree with the experimental data. Furthermore, the first principle simulation showed that the binding energies between dye and SWCNTs are heightened as the carbon nanotube diameter gets bigger because of the increase of π–π interaction.

Rajabia et al. [78] reported the efficient removal of malachite green from its aqueous solution by MWCNTs functionalized with the carboxylate group. This nanomaterial showed a fast removal of malachite green dye, being 10 min the optimum contact time. In addition, the MWCNT-COOH presented an interesting adsorption performance with a maximum adsorption capacity (in 10 min) of 11.77 mg g^{-1} at 328 K and pH 9. The adsorption equilibrium and kinetic data were well fitted and found to be in good agreement with the Langmuir isotherm and the intraparticle diffusion kinetic models, respectively. The authors found that the adsorption capacity of malachite green increased with increasing contact time, temperature, and pH of the working solution. The authors affirm also that the carboxylate groups are the key to success for the adsorption of cationic dye on MWCNT. The negative surface charge increases the solubility of the MWCNT, in addition to increasing the electrostatic interaction between the CNT and malachite green.

Single-walled and multi-walled carbon nanotubes were used as adsorbents for the removal of Reactive Blue 4 dye from aqueous effluents [25]. The contact time to obtain equilibrium isotherms was fixed at 4 h at 298–323 K for both CNTs. The general-order kinetic equation provided the best fit to the adsorption data for the both nanotubes. The equilibrium data, for all temperatures, were well described by the Liu isotherm model, and the maximum sorption capacities for adsorption (at 323 K) of dye were 567.7 and 502.5 mg g^{-1} for SWCNT and MWCNT, respectively. The textural properties CNTs justified the higher adsorption capacity of SWCNTs (13.0% higher than the value obtained for MWCNT). This has the total pore volume and specific surface area 91.9 and 114.3%, respectively, higher than those of multi-walled. The change in enthalpy of adsorption between CNT—dye indicated that adsorption was an endothermic process. The interaction of (8, 0) SWCNTs Reactive Blue 4 dye was investigated using ab initio calculations based on density functional theory (DFT). The results from first principles calculations showed that the (8, 0) SWCNT interacts with dye the textile through an electrostatic interaction. These results agree with the experimental data.

MWCNT functionalized with thiol group (with the different percentage of cysteamine mass) was used as adsorbent for the rapid removal of Methylene Blue (MB) dye from the liquid phase [79]. The authors found that the adsorption capacity increases with increase in the initial concentrations of MB dye, from 10 to 40 mg L^{-1}. Furthermore, it was concluded that the adsorption capacity of MB dye on MWCNT functionalized thiol increased with increasing the temperature, from 283 to 303 K.

Jauris et al. [80] from ab initio simulation based on the DFT, evaluated the interaction between (5, 5) and (8, 0) SWCNTs with and without vacancy and two dyes, MB and Acridine Orange. Additionally, the authors evaluated the influence of the nanotube diameter in the adsorption process. The results presented that the main configurations show an average binding energy 0.75 eV, a characteristic physisorption. Likewise, the SWCNT with one vacancy current lower binding energies than the SWCNT without vacancy, excluding the case of MB plus SWCNT. Also, the dyes plus (5, 5) SWCNT present superior binding energies than dyes plus (8, 0) SWCNT. Dyes plus (5, 5) and (8, 0) SWCNTs current lower binding energies compared to respective dyes plus (16, 0) and (25, 0) SWCNT, because of the increase of π–π interaction with increase diameter tube. Those data are very promising since they propose that CNTs are appropriate for real textile effluent treatment.

More recently, the graphene family, especially the graphene oxide and the reduced graphene oxide, have also emerged as a promising family of adsorbent for the removal of various dyes from waste effluents. There are several current studies reporting the application of that family with adsorbent [81].

Peng et al. [82] verified the capacity of adsorption of MB dye on GO, synthetized from an economical and resourceful amorphous graphite. They found that the C/O mass ratio of GO reached 1.84 and that thin layers accounted for 83.76%. The results of batch adsorption showed that the interaction between GO and MB dye in the presence of cations decreased in the sequence $Na^+ \approx Li^+ > K^+$, while it was reduced in the order $ClO_4^- > NO_4^- > Cl^-$ in the presence of anions. The removal of MB in ClO_4^- was independent of solution pH, which may be credited to the synergistic effect between ClO_4^- and GO. The kinetic of adsorption process was well described by pseudo-second-order equation and the adsorption isotherm good agreement with the Langmuir model. The maximum adsorption capacity of MB on GO was 2273.6 mg g^{-1} at 298 K.

Yan et al. [83] evaluated too the influence of oxidation degree of GO on the adsorption of MB dye from aqueous solutions. The GO shows a fast and pH-independent MB adsorption. Additionally, the dye uptakes of GO exponentially increase with the increase of oxidation degree, and the adsorption behavior of GO would change from a Freundlich to a Langmuir adsorption models as the oxidation degree increases. This can be credited for enhanced exfoliation degree of the carbon planes in graphite caused by oxidation and the appearance of new and more active adsorption sites. Also, it was found that the adsorption mechanism can change of hydrophobic π–π for electrostatic interactions with the oxidation degree of GO. For the case where the oxidation degree was low, the dye interacted with the GO due primarily through parallel π–π stacking interactions and form multilayer adsorption; while GO with higher oxidation degree dominantly shows monolayer adsorption by electrostatic interactions. It is this last most efficient mechanism and the maximum sorption capacities for adsorption of MB dye was 598.8 mg g^{-1} for the GO with higher with oxidation degree. Moreover, it is also found that the GO could be regenerated and reused for a few cycles.

Padhi et al. [84] evaluating adsorption capacity of the Congo Red anionic dye by GO in the presence of MB cationic dye. These authors found that the GO increases your adsorption capacity in 96% toward Congo Red in presence of MB, at pH = 2. This intensely increased of Congo Red adsorption on GO in presence of MB dye which can be probably clarified by: charge density modification of GO by dyes from high negative charge density to positive charge density for easing the contact toward the anionic dye molecule; possible electrostatic interaction arising between the GO and the MB, Congo Red, and MB + Congo Red; and the existence of robust electrostatic attraction between the existing pH-dependent chemical structure of Congo Red and MB synthetic dyes in aqueous solution. The thermodynamic parameters such as Gibbs energy change show that the adsorption process of MB + Congo Red dyes by GO is spontaneous.

Kim et al. [40] demonstrate adsorption behaviors of MB and Acid Red I dyes on three-dimensional (3D) rGO macrostructure. The distinct interactions of two dyes with the active sites result in the different adsorption behaviors, thus represent the notable disparity in the adsorption capacities and kinetics. The equilibrium data for MB are fitted to Langmuir isotherm model, while Freundlich model is suitable for the equilibrium isotherm of Acid Red I. The adsorption rates of both dyes are found to follow the pseudo-second-order kinetic model. The 3D rGO macrostructures are more favorable for the adsorption of cationic dyes rather than the anionic dyes due to strong specific interactions.

Robati et al. [85] reported the fast adsorption of the malachite green dye from the liquid phase onto the rGO and GO surfaces. Based on batch adsorption experiment, the authors found that the adsorption process this dye by nanoadsorbents is influenced by parameters such as contact time, temperature, and initial pH. It was found that by increasing the pH (from 3 to 9), the removal of dye by the GO and rGO adsorbent was decreased. Moreover, the removal of malachite green dye by rGO and GO nanoadsorbents was increased by increasing the temperature, which directly designates toward the endothermic nature of the adsorption process. The adsorption kinetic data of malachite green dye on both adsorbents was found to be well fitted by the pseudo-first-order kinetic model.

4.2 Adsorption of Emerging Contaminants

Over recent decades, emerging contaminants (EC) such as pharmaceuticals and by-products of human activity, personal care products, fire retarding agents and polymer additives, pesticides and herbicides, disinfectants among others have become a concern in water and wastewater treatment plants [2, 81, 86]. Most conventional wastewater treatment processes are not planned for eliminating these contaminants [86]. Despite half-lives persist for a short period, EC exhibit pseudo-persistent phenomenon since of recurrent uses in routine life [86, 87]. These can have bio-accumulative and damaging effects on the survival of aquatic organisms, fauna, flora, as well as to human health [13, 86, 87].

Numerous studies have explored the application potential of CNA for the treatment of wastewater contaminated with EC. CNA have recently been considered emerging adsorbents for this application since it has developed pore structure, large specific surface area, and good chemical and mechanical strength. Given the background, some batch adsorption and studies on Theoretical Investigation have been directed to evaluate the EC adsorption behavior on CNA.

MWCNTs modified with HNO_3 were used as nanoadsorbents for removal of diclofenac [86]. The best operating conditions were determined as drug initial concentration 50 mg L^{-1}, adsorbent dosage 270 mg, temperature 298.15 K, contact time 60 min, and pH 7.0. The MWCNTs modified by HNO_3 in the range of 10–30 nm displayed a best adsorption removal efficiency of about 95%. The kinetic study shows that the process is controlled by multiple steps and adsorption of the pharmaceutical can be well described by the pseudo-second-order kinetic equation. The equilibrium data were well described by the Freundlich isotherm equation, and the maximum sorption capacity for adsorption was 24 mg g^{-1} at 298.15 K. The authors found that the diclofenac adsorption capacity decreases with increasing temperature. The thermodynamic parameters displayed that the adsorption of the drug by CNTs is exothermic and spontaneous.

The adsorption of two antibiotics, sulfamethoxazole and lincomycine by SWCNT and MWCNT, was investigated using batch adsorption [88]. The adsorption results were compared with those of activated carbon (AC). Adsorption isotherms of the two antibiotics on different adsorbents were nonlinear and described by Freundlich isotherm equation. The adsorption capacity following order: SWCNT > AC > MWCNT. The authors attribute the low adsorption capacity of two antibiotics by MWCNT to the lower specific surface area of this CNT, compared to other adsorbents. Since the two studied drugs are very hydrophilic, the adsorption mechanism was attributed to hydrogen bonds and electrostatic interactions.

SWCNT pristine and functionalized with hydroxy and carboxyl groups were used with adsorbents for removing to six kinds of chlorophenols (intermediates employed to produce pesticides, biocides, and dyes) [89]. The adsorption equilibrium was obtained in 2 h, and the adsorption kinetics was well described by pseudo-second-order model. According to the authors, every adsorption isotherms were well fitted with Polanyi–Manes, Freundlich, and Langmuir models. The adsorption capacity of chlorophenols by SWCNTs reduced as the following order: pristine SWCNT > SWCNT-OH > SWCNT-COOH, demonstrating that oxidation of SWCNT surface prejudices significantly the adsorption capacity. The adsorption capacity of SWCNT was weakened by anchoring of oxygen-containing functional groups on the surface, by the increase of hydrophilicity and the reduction of interaction, as well as to the loss of specific surface area. The adsorption capacity of pristine SWCNT, SWCNT-COOH, and SWCNT-OH for chlorophenols diverse, respectively, from 19 to 84 mg g^{-1}, from 17 to 65 mg g^{-1}, and from 19 to 65 mg g^{-1}. The authors compared the chlorophenols adsorption capacity by different adsorbents, i.e., SWCNT, MWCNT, and AC. The studies indicating that the adsorption rate of CNT was much faster than that of AC.

The adsorption of endocrine disrupting compounds (EDC) Bisphenol A and 7α-ethinyl estradiol, from brackish water, reproduction seawater or the combination thereof using SWCNT as nanoadsorbent was measured [90]. Linearized Langmuir and Freundlich equations were used to fitting the isothermal data of the adsorption of Bisphenol A and 7α-ethinyl estradiol by SWCNT in the three kinds of water sources. The data show higher removal capacity for 17α-ethinyl estradiol (from 95 to 98%) than for Bisphenol A (from 75 to 80%). Variations in the water chemistry conditions of the water sources did not significantly influence the overall adsorption of Bisphenol A and 7α-ethinyl estradiol. Nevertheless, varying the pH of the water sources from 3.5 to 11 displayed a reduction in the removal of Bisphenol A, but it did not affect the adsorption of 7α-ethinyl estradiol. The authors found that the hydrophobic interactions between SWCNTs and EDCs maybe the leading adsorption mechanism. Nonetheless, π–π electron donor–acceptor interactions can be another possible adsorption mechanism of EDCs.

In recent times, the graphene family has attracted scientists and engineers due to its unique properties, and low production costs when to compared with other carbon materials [5, 81]. With the increasing demand for clean drinking water, this family characterized by large delocalized pi (π) electrons, extremely high surface area, and tunable chemical properties make them extremely compelling for use as adsorbents for environmental pollution cleanup [39–48, 60, 61, 81], especially drink water contaminated with EC.

Recently, the interactions of sodium diclofenac with different graphene kinds were investigated using first principles calculations based on DFT [13]. In addition, the authors compared the theoretical results with data obtained experimentally. Through batch adsorption experiments, it was found that rGO was a good nanoadsorbent for extracting of sodium diclofenac from reproduction wastewater. General-order kinetic adsorption model demonstrated the best fit to the experimental data. The equilibrium data were fitted by Liu isotherm equation. Maximum sorption capacity for adsorption of the diclofenac was 59.67 mg g^{-1}. The diclofenac drug adsorption onto pristine graphene, graphene with a vacancy, rGO and functionalized graphene nanoribbons were simulated by ab initio and provided a good understanding of the adsorption process of diclofenac molecule on graphene family. The results indicate a physical adsorption regime in all cases. To pristine graphene and graphene with a single vacancy, the most stable configurations are those ones where the sodium diclofenac present hexagonal carbon rings over graphene rings forming a Bernal-like stacking, thus demonstrating the prevalence of π–π interactions. Regarding the adsorption of the drug onto nanoribbons or functionalized graphene, binding energies to increase the number of functional groups increase on the surface. In addition, the most pertinent interaction in terms of binding energy can be credited to hydrogen bonding. Founded on these results, the first principles calculations and the batch adsorption experiments point out that the graphene family is attractive materials for extracting sodium diclofenac from aqueous effluents.

Nam et al. [51] also evaluated, by experimental and theoretical investigations, the adsorption capacity of diclofenac by graphene family. However, these researchers used the GO with nanoadsorbent. In addition, the authors evaluated the capacity of adsorption of the sulfamethoxazole onto GO. The adsorption performance of pharmaceutical was explored in terms of contact time, pH solution, and GO dosage. Both pharmaceuticals compound was well fitted by the Freundlich isotherm model. The GO removed up to 12 and 35% of sulfamethoxazole and diclofenac, respectively, within 6 h, and an increase in adsorbent dosage enhanced the adsorption of diclofenac. The main interacting mechanisms were predicted to be π–π electron donor-acceptor and hydrophobic interactions between the both drugs and GO. The molecular modeling proposed that the adsorption site of diclofenac is more exposed to the GO surface, which results in a greater adsorption capacity when compared to the other pharmaceutical compound. The binding energy of diclofenac (-18.8 kcal mol^{-1}) had a more favorable on the GO surface than sulfamethoxazole (15.9 kcal mol^{-1}). Furthermore, the effects of sonication on adsorption process were examined by authors. That check that the sonication expressively improved the removal of sulfamethoxazole and diclofenac, 30 and 75%, respectively, due to dispersion of GO sheet and the reduction of oxygen-containing functional groups on the GO surface.

The ability of GO to adsorption of sulfamethoxazole was also evaluated by Chen et al. [91]. In addition, these authors investigate the capacity of GO to remove the ciprofloxacin from aqueous solutions. The ciprofloxacin isotherm fitted slightly better with the Freundlich model. On the other hand, both the Freundlich and Langmuir models fitted the sulfamethoxazole isotherm well. The maximum adsorption capacity of sulfamethoxazole and ciprofloxacin by GO were, respectively, of 240 and 379 mg g^{-1}. According to the authors, the adsorption of sulfamethoxazole was mostly though π–π electron donor-acceptor attraction on the basal planes of the GO; while ciprofloxacin adsorption was mostly controlled by the electrostatic attraction. Solution pH showed a strong effect on the sorption ability of GO to the two antibiotics. At pH = 2, the GO adsorption capacity decreased for both drugs, and at pH = 9 the GO completely lost sulfamethoxazole adsorption capacity, but still showed strong sorption to ciprofloxacin. For the case of sulfamethoxazole, the effect of $CaCl_2$ and NaCl on adsorption was weaker, but higher ionic strength also reduced the adsorption this drug by GO. Both $CaCl_2$ and NaCl decreased the adsorption ability of ciprofloxacin by GO.

Rostamian and Behnejad [92] also investigated the removal of sulfamethoxazole by CAN. These authors used graphene nanosheet (GNS) and graphene oxide nanosheet (GOS) with specific surface area of 106 and 123 $m^2\ g^{-1}$, respectively, as adsorbents. Five factors were determinant in the adsorption of antibiotic: temperature, solution pH, initial antibiotic concentration, the amount of adsorbent, and contact time. The results displayed that initial pH $\sim$ 6, the adsorbent dosage of 0.010 g, contact time $\sim$ 110 min at 25 °C were optimum for both systems. The authors observed that the interaction between the antibiotic and the two adsorbents decreases with the increase in temperature from 25 to 45 °C, consequently decreasing the maximum sulfamethoxazole adsorption capacity. The Koble–Corrigan and Redlich–Peterson isotherm models represented the equilibrium adsorption data of sulfamethoxazole, while kinetic adsorption data were well fitted by pseudo-second-order equation on both nanoadsorbent. Through of the thermodynamic calculations, the authors demonstrated that the adsorption process was exothermic and spontaneous for all temperatures. However, the adsorption process of sulfamethoxazole onto GNS and GOS becomes less favorable at higher temperatures. In addition, through of change in enthalpy, the authors determined the interaction of the antibiotic with both CNA was dominated by electrostatic attraction [25].

Dong et al. [93] evaluated the potential of application of GO directly a filter for the removal of levofloxacin from aqueous solution. Batch and fixed-bed experiments were showed to determine the sorption behaviors of antibiotic onto the GO. In the batch experimental, GO showed high adsorption of the drug with Langmuir maximum adsorption capacity of 256.6 $mg\ g^{-1}$. The removal of antibiotic by GO in fixed-bed columns was strong under all tested conditions. The removal efficiency of the levofloxacin in the GO columns increased with increasing the amount of GO, nonetheless, decreased with increasing injection flow rate. The fixed-bed experimental results were well described by the Bed Depth Service Time model, indicating the model can be employed for the design of GO filters in large-scale applications.

Kwon and Lee [59] reported the use of rGO as adsorbent for the removal of Bisphenol A from aqueous solutions. The authors also evaluated the influence of the reduction method used in the synthesis of rGO in the adsorption of this EDC. The nanoadsorbents were prepared by a thermal exfoliation method (T-rGO) and a chemical reduction method (H-rGO) using hydrazine. The specific surface area (BET) and pore volume of T-rGO and H-rGO were 494 $m^2\ g^{-1}$ and 1.56 $cm^3\ g^{-1}$, and 139 $m^2\ g^{-1}$ and 0.15 $cm^3\ g^{-1}$, respectively. The adsorption kinetic data were fitted well with the pseudo-second-order-model and the T-rGO displayed an adsorption kinetic constant 200 times larger than that of H-rGO. The authors attributed the fast adsorption rates and selective adsorption characteristics to the synergistic effect of the hydrophilic surface functional groups and the hydrophobic graphene layers of T-rGO. The higher concentrations of hydroxyl groups and other oxygen functional groups on the T-rGO surface to favor the both π–π interactions and hydrogen bonds with Bisphenol A. Equilibrium adsorption experiments were carried out using both rGOs, and the results were fitted to Langmuir and Freundlich adsorption isotherms models. The adsorption data of both Bisphenol A on T-rGO were well described using the Freundlich isotherm, in contrast, the adsorption data of H-rGO were better fitted to the Langmuir isotherm model. The Langmuir model showed that the maximum adsorption capacity of T-rGO and H-rGO toward Bisphenol A was 96.2 and 81.3 mg g^{-1}, respectively. The thermodynamic parameters of Bisphenol A adsorption on T-rGO indicated that the adsorption is physical and exothermic.

Bele et al. [94] evaluated the influence on the reduction degree of GO in the adsorption of Bisphenol A. The authors verified that the adsorption capacities this EC were increased with increasing the reduction degree of GO. The Langmuir isotherm model better described the adsorption of Bisphenol A onto all rGOs and the maximum adsorption capacity was 94.06 mg g^{-1}. The increase in the degree of GO reduction reduced the amount of oxygen-containing functional groups on the surface of reduced samples, resulting in the increase of the π–π electron donor–acceptor interaction between sorbent–adsorbate and to a linear increase of adsorption capacity. The adsorption of Bisphenol A followed pseudo-second-order kinetics model. The thermodynamic analysis indicated that it was spontaneous and endothermic.

5 Conclusion

In this chapter covered a wide range of studies about adsorption of carbon nanomaterials as nanoadsorbents, which have been used so far for removal of numerous organic contaminants from the water and effluent. Furthermore, a discussion of the main kinetics and isotherms models used to obtain information on the mechanisms and dynamics of the process was carried. The high simplicity, efficiency, and ease in the scaling-up of adsorption processes employing carbon nanomaterials make the adsorption technique attractive for removal of organic compounds. Most of the studies stated in the literature have focused on the synthesis of carbon nanomaterials from low-cost resources compared with the performance of readily accessible commercial activated carbons. As demonstrated, carbon nanomaterials can play a significant role in this context and, consequently, an emerging area of research is the development of new materials with a high capacity for adsorption process.

Acknowledgements The authors acknowledge funding from Brazilian agencies CNPq and CAPES.

References

1. Duncan, J., N. Savage, A. Street, and Sustich, R. 2014. *Nanotechnology applications for clean water: Solutions for improving water quality*. 2nd ed. Norwich, NY: Micro & Nano Technologies, William Andrew Inc.
2. Petrie, B., R. Ruth, and B. Kasprzyk-Hordern. 2015. A review on emerging contaminants in wastewaters and the environment: Current knowledge, understudied areas and recommendations for future monitoring. *Water Research* 72: 3–27.
3. Kümmerer, K. 2008. *Pharmaceuticals in the environment: Sources, fate, effects and risks*. 3rd ed. Berlin: Springer.
4. Barceló, D. 2012. *Emerging organic contaminants and human health*. Berlin: Springer.
5. Bergmann, C.P., and F.M. Machado. 2015. *Carbon nanomaterials as adsorbents for environmental and biological applications*. New York: Springer International Publishing.
6. Dąbrowski, A. 2001. Adsorption—from theory to practice. *Advances in Colloid and Interface Science* 93: 135–224.
7. Ruthven, D.M. 1984. *Principles of adsorption and adsorption processes*. New York: Wiley.
8. Gupta, V.K., and T.A. Saleh. 2013. Sorption of pollutants by porous carbon, carbon nanotubes and fullerene—An overview. *Environmental Science and Pollution Research* 20: 2828–2843.
9. Reis, G.S., M. Wilhelm, T.C.A. Silva, K. Rezwan, C.H. Sampaio, E.C. Lima, and S.M.A.G. U. de Souza. 2016. The use of design of experiments for the evaluation of the production of surface rich activated carbon from sewage sludge via microwave and conventional pyrolysis. *Applied Thermal Engineering* 93: 590–597.
10. Machado, F.M., C.P. Bergmann, T.H.M. Fernandes, et al. 2011. Adsorption of Reactive Red M-2BE dye from water solutions by multi-walled carbon nanotubes and activated carbon. *Journal of Hazardous Materials* 192: 1122–1131.
11. Prola, L.D.T., F.M. Machado, C.P. Bergmann, et al. 2013. Adsorption of Direct Blue 53 dye from aqueous solutions by multi-walled carbon nanotubes and activated carbon. *Journal of Environmental Management* 130: 166–175.
12. Patiño, Y., E. Díaz, S. Ordóñez, E. Gallegos-Suarez, A. Guerrero-Ruiz, and I. Rodríguez-Ramos. 2015. Adsorption of emerging pollutants on functionalized multiwall carbon nanotubes. *Chemosphere* 136: 174–180.
13. Jauris, I.M., C.F. Matos, C. Saucier, et al. 2016. Adsorption of sodium diclofenac on graphene: A combined experimental and theoretical study. *Physical Chemistry Chemical Physics* 18: 1526–1536.
14. Machado, F.M., S.A. Carmalin, E.C. Lima, et al. 2016. Adsorption of Alizarin Red S dye by carbon nanotubes: An experimental and theoretical investigation. *Journal of Physical Chemistry C* 120: 18296–18306.
15. Bai, H., X. Zan, L. Zhang, and D.D. Sun. 2015. Multi-functional CNT/ZnO/TiO_2 nanocomposite membrane for concurrent filtration and photocatalytic degradation. *Separation and Purification Technology* 156: 922–930.
16. Kim, J.D., H. Yun, G.C. Kim, C.W. Lee, and H.C. Choi. 2013. Antibacterial activity and reusability of CNT-Ag and GO-Ag nanocomposites. *Applied Surface Science* 283: 227–233.
17. Gupta, V.K., S. Agarwal, and T.A. Saleh. 2011. Synthesis and characterization of alumina-coated carbon nanotubes and their application for lead removal. *Journal of Hazardous Materials* 85 (1): 17–23.
18. Guo, J., R. Wang, W.W. Tjiu, J. Pan, and T. Liu. 2012. Synthesis of Fe nanoparticles@-graphene composites for environmental applications. *Journal of Hazardous Materials* 225: 63–73.
19. Terrones, M., A.R. Botello-Méndez, J. Campos-Delgado, et al. 2010. Graphene and graphite nanoribbons: Morphology, properties, synthesis, defects and applications. *Nano Today* 5 (4): 351–372.

20. O'Connell, J.M. 2006. *Carbon nanotubes: Properties and applications*. LLC, New York: Taylor & Francis Group.
21. Novoselov, K.S., A.K. Geim, S.V. Morozov, et al. 2004. Electric field effect in atomically thin carbon films. *Science* 306: 666–669.
22. Iijima, S. 1991. Helical microtubules of graphitic carbon. *Nature* 354: 56–58.
23. Bethune, D.S., C.H. Klang, M.S. De Vries, et al. 2003. Cobalt-catalysed growth of carbon nanotubes with single-atomic-layer walls. *Nature* 363: 605–607.
24. Iijima, S., and T. Ichihashi. 1993. Single-shell carbon nanotubes of 1-nm diameter. *Nature* 363: 603–605.
25. Machado, F.M., C.P. Bergmann, E.C. Lima, et al. 2012. Adsorption of Reactive Blue 4 dye from water solutions by carbon nanotubes: Experiment and theory. *Physical Chemistry Chemical Physics* 14: 11139–11153.
26. Babaa, M.R., N. Dupont-Pavlovsky, E. McRae, and K. Masenelli-Varlot. 2004. Physical adsorption of carbon tetrachloride on as-produced and on mechanically opened single walled carbon nanotubes. *Carbon* 42: 1549–1554.
27. Ren, X., C. Chen, M. Nagatsu, and X. Wang. 2011. Carbon nanotubes as adsorbents in environmental pollution management: A review. *Chemical Engineering Journal* 170: 395–410.
28. Upadhyayula, V.K.K., S. Deng, M.C. Mitchell, and G.B. Smith. 2009. Application of carbon nanotube technology for removal of contaminants in drinking water: A review. *Science of the Total Environment* 408: 1–13.
29. Sze, M.F.F., V.K.C. Lee, and G. McKay. 2008. Simplified fixed bed column model for adsorption of organic pollutants using tapered activated carbon columns. *Desalination* 218: 323–333.
30. Yuan, W., B. Li, and L. Li. 2011. A green synthetic approach to graphene nanosheets for hydrogen adsorption. *Applied Surface Science* 257: 10183–10187.
31. Negishi, R., H. Hirano, Y. Ohno, K. Maehashi, K. Matsumoto, and Y. Kobayashi. 2011. Layer-by-layer growth of graphene layers on graphene substrates by chemical vapor deposition. *Thin Solid Films* 519: 6447–6452.
32. Brodie, B.C. 1859. On the atomic weight of graphite. *Philosophical Transactions of the Royal Society of London, Series A* 149: 249–259.
33. Staudenmaier, L. 1898. Verfahren zur darstellung der graphitsaure. *Berichte der Deutschen Chemischen Gesellschaft* 31: 1481–1487.
34. Hummers, W.S., and R.E. Offeman. 1958. Preparation of graphitic oxide. *Journal of the American Chemical Society* 80: 1339.
35. Bianco, A., H.-M. Cheng, T. Enoki, et al. 2013. All in the graphene family—A recommended nomenclature for two-dimensional carbon materials. *Carbon* 65: 1–6.
36. Hernandez, Y., V. Nicolosi, M. Lotya, et al. 2008. High-yield production of graphene by liquid-phase exfoliation of graphite. *Nature Nanotechnology* 3: 563–568.
37. Sun, Y., Q. Wu, and G. Shi. 2011. Graphene based new energy materials. *Energy & Environmental Science* 4: 1113–1132.
38. Stankovich, S., D.A. Dikin, G.H.B. Dommett, et al. 2006. Graphene-based composite materials. *Nature* 442: 282–286.
39. Xiao, J., W. Lv, Z. Xie, Y. Tan, Y. Song, and Q. Zheng. 2016. Environmentally friendly reduced graphene oxide as a broad-spectrum adsorbent for anionic and cationic dyes via π–π interactions. *J Mater Chem A* 4: 12126–12135.
40. Kim, H., S.-O. Kang, S. Park, and H.S. Park. 2015. Adsorption isotherms and kinetics of cationic and anionic dyes on three-dimensional reduced graphene oxide macrostructure. *Journal of Industrial and Engineering Chemistry* 21: 1191–1196.
41. Tsoufis, T., G. Tuci, S. Caporali, D. Gournis, and G. Giambastiani. 2013. *p*-Xylylenediamine intercalation of graphene oxide for the production of stitched nanostructures with a tailored interlayer spacing. *Carbon* 59: 100–108.
42. Li, Y., Q. Du, J. Wang, et al. 2013. Defluoridation from aqueous solution by manganese oxide coated graphene oxide. *Journal of Fluorine Chemistry* 148: 67–73.

43. Lee, Y.-C., S.-J. Chang, M.-H. Choi, T.-J. Jeon, T. Ryu, and Y.S. Huh. 2013. Self-assembled graphene oxide with organo-building blocks of Fe-aminoclay for heterogeneous Fenton-like reaction at near-neutral pH A batch experiment. *Applied Catalysis B: Environmental* 142–143: 494–503.
44. Sharma, P., and M.R. Das. 2013. Removal of a cationic dye from aqueous solution using graphene oxide nanosheets: Investigation of adsorption parameters. *Journal of Chemical and Engineering Data* 58: 151–158.
45. Mishra, A.K., and S. Ramaprabhu. 2011. Removal of metals from aqueous solution and sea water by functionalized graphite nanoplatelets based electrodes. *Journal of Hazardous Materials* 185: 322–328.
46. Ren, X., J. Li, X. Tan, and X. Wang. 2013. Comparative study of graphene oxide, activated carbon and carbon nanotubes as adsorbents for copper decontamination. *Dalton Transactions* 42: 5266–5274.
47. Lingamdinne, L.P., J.R. Koduru, H. Roh, Y.-L. Choi, Y.-Y. Chang, and J.-K. Yang. 2016. Adsorption removal of Co(II) from waste-water using graphene oxide. *Hydrometallurgy* 165: 90–96.
48. Yang, S.-T., S. Chen, Y. Chang, A. Cao, Y. Liu, and H. Wang. 2011. Removal of methylene blue from aqueous solution by graphene oxide. *Journal of Colloid and Interface Science* 359: 24–29.
49. Xu, J., and Y.-F. Zhu. 2013. Elimination of Bisphenol A from water via graphene oxide adsorption. *Acta Physico-Chimica Sinica* 29: 829–836.
50. Gao, Y., L. Zhang, H. Huang, J. Hu, S. Shah, and X. Su. 2012. Adsorption and removal of tetracycline antibiotics from aqueous solution by graphene oxide. *Journal of Colloid and Interface Science* 368: 540–546.
51. Nam, S.-W., C. Jung, H. Li, et al. 2015. Adsorption characteristics of diclofenac and sulfamethoxazole to graphene oxide in aqueous solution. *Chemosphere* 136: 20–26.
52. Pei, S., and H.M. Cheng. 2012. The reduction of graphene oxide. *Carbon* 50: 3210–3228.
53. Park, S., J. An, J.R. Potts, A. Velamakanni, S. Murali, and R.S. Ruoff. 2011. Hydrazine-reduction of graphite and graphene oxide. *Carbon* 49: 3019–3023.
54. Bai, Y., R.B. Rakhi, W. Chen, and H.N. Alshareef. 2013. Effect of pH-induced chemical modification of hydrothermally reduced graphene oxide on supercapacitor performance. *Journal of Power Sources* 233: 313–319.
55. Song, S., H. Yang, C. Su, Z. Jiang, and Z. Lu. 2016. Ultrasonic-microwave assisted synthesis of stable reduced graphene oxide modified melamine foam with superhydrophobicity and high oil adsorption capacities. *Chemical Engineering Journal* 306: 504–511.
56. Fei, P., Q. Wang, M. Zhong, and B. Su. 2016. Preparation and adsorption properties of enhanced magnetic zinc ferrite-reduced graphene oxide nanocomposites via a facile one-pot solvothermal method. *Journal of Alloys and Compounds* 635: 411–417.
57. Liu, F.-F., J. Zhao, S. Wang, and B. Xing. 2016. Adsorption of sulfonamides on reduced graphene oxides as affected by pH and dissolved organic matter. *Environmental Pollution* 210: 85–93.
58. Bi, H., X. Xie, K. Yin, et al. 2012. Spongy graphene as a highly efficient and recyclable sorbent for oils and organic solvents. *Advanced Functional Materials* 22: 4421–4425.
59. Kwon, J., and B. Lee. 2015. Bisphenol A adsorption using reduced graphene oxide prepared by physical and chemical reduction methods. *Chemical Engineering Research and Design* 104: 519–529.
60. Wang, J., and B. Chen. 2015. Adsorption and coadsorption of organic pollutants and a heavy metal by graphene oxide and reduced graphene materials. *Chemical Engineering Journal* 281: 379–388.
61. Bai, S., X. Shen, G. Zhu, et al. 2013. The influence of wrinkling in reduced graphene oxide on their adsorption and catalytic properties. *Carbon* 60: 157–168.
62. Largegren, S. 1898. About the theory of so-called adsorption of soluble substances. *Kungliga Suensk Vetenskapsakademiens Handlingar* 24: 1–39.

63. Ho, Y.S. 2006. Review of second-order models for adsorption systems. *Journal of Hazardous Materials* 136: 681–689.
64. Alencar, W.S., E.C. Lima, B. Royer, et al. 2012. Application of aqai stalks as biosorbents for the removal of the dye Procion Blue MX-R from aqueous solution. *Separation Science and Technology* 47: 513–526.
65. Lopes, E.C.N., F.S.C. dos Anjos, E.F.S. Vieira, and A.R. Cestari. 2003. An alternative Avrami equation to evaluate kinetic parameters of the interaction of Hg(II) with thin chitosan membranes. *Journal of Colloid and Interface Science* 263: 542–547.
66. Vaghetti, J.C.P., E.C. Lima, B. Royer, et al. 2009. Pecan nutshell as biosorbent to remove Cu (II), Mn(II) and Pb(II) from aqueous solutions. *Journal of Hazardous Materials* 162: 270–280.
67. Langmuir, I. 1918. The adsorption of gases on plane surfaces of glass, mica and platinum. *Journal of the American Chemical Society* 40: 1361–1403.
68. Freundlich, H. 1906. Adsorption in solution. *Physical Chemistry Society* 40: 1361–1368.
69. Sips, R. 1948. On the structure of a catalyst surface. *The Journal of Chemical Physics* 16: 490–495.
70. Liu, Y., H. Xu, S.F. Yang, and J.H. Tay. 2003. A general model for biosorption of Cd^{2+}, Cu^{2+} and Zn^{2+} by aerobic granules. *Journal of Biotechnology* 102: 233–239.
71. Redlich, O., and D.L. Peterson. 1959. A useful adsorption isotherm. *Journal of Physical Chemistry* 63: 1024–1027.
72. Lima, E.C., M.A. Adebayo, and F.M. Machado. 2015. Chapter 3—Kinetic and equilibrium models of adsorption. In *Carbon nanomaterials as adsorbents for environmental and biological applications*, ed. C.P. Bergmann, and F.M. Machado, 33–69. Berlin: Springer.
73. Levenspiel, O. 1999. *Chemical Reaction Engineering*, 3rd ed, 1999. New York: Wiley.
74. Ribas, M.C., M.A. Adebayo, L.D.T. Prola, et al. 2014. Comparison of a homemade cocoa shell activated carbon with commercial activated carbon for the removal of reactive violet 5 dye from aqueous solutions. *Chemical Engineering Journal* 248: 315–326.
75. Vaghetti, J.C.P., E.C. Lima, B. Royer, et al. 2008. Application of Brazilian-pine fruit coat as a biosorbent to removal of Cr(VI) from aqueous solution. Kinetics and equilibrium study. *Biochemical Engineering Journal* 42: 67–76.
76. Lima, E.C., A.R. Cestari, and M.A. Adebayo. 2016. Comments on the paper: A critical review of the applicability of Avrami fractional kinetic equation in adsorption-based water treatment studies. *Desalination and Water Treatment* 57: 19566–19571.
77. Thue, P.S., E.C. Lima, J.M. Sieliechi, et al. 2017. Effects of first-row transition metals and impregnation ratios on the physicochemical properties of microwave-assisted activated carbons from wood biomass. *Journal of Colloid and Interface Science* 486: 163–175.
78. Rajabia, M., B. Mirzab, K. Mahanpoorc, et al. 2016. Adsorption of malachite green from aqueous solution by carboxylate group functionalized multi-walled carbon nanotubes: Determination of equilibrium and kinetics parameters. *Journal of Industrial and Engineering Chemistry* 34: 130–138.
79. Robati, D., B. Mirza, R. Ghazisaeidi, et al. 2016. Adsorption behavior of methylene blue dye on nanocomposite multi-walled carbon nanotube functionalized thiol (MWCNT-SH) as new adsorbent. *Journal of Molecular Liquids* 216: 830–835.
80. Jauris, I.M., S.B. Fagan, M.A. Adebayo, and F.M. Machado. 2016. Adsorption of acridine orange and methylene blue synthetic dyes and anthracene on single wall carbon nanotubes: A first principle approach. *Computational and Theoretical Chemistry* 1076: 42–50.
81. Chowdhury, S., and R. Balasubramanian. 2014. Recent advances in the use of graphene-family nanoadsorbents for removal of toxic pollutants from wastewater. *Advances in Colloid and Interface Science* 204: 35–56.
82. Peng, W., H. Li, Y. Liu, and S. Song. 2016. Adsorption of methylene blue on graphene oxide prepared from amorphous graphite: Effects of pH and foreign ions. *Journal of Molecular Liquids* 221: 82–87.
83. Yan, H., X. Tao, Z. Yang, et al. 2014. Effects of the oxidation degree of graphene oxide on the adsorption of methylene blue. *Journal of Hazardous Materials* 268: 191–198.

84. Padhia, D.K., K.M. Parida, and S.K. Singh. 2016. Mechanistic aspects of enhanced congo red adsorption over graphene oxide in presence of methylene blue. *Journal of Environmental Chemical Engineering* 4: 3498–3511.
85. Robati, D., M. Rajabi, O. Moradi, et al. 2016. Kinetics and thermodynamics of malachite green dye adsorption from aqueous solutions on graphene oxide and reduced graphene oxide. *Journal of Molecular Liquids* 214: 259–263.
86. Hu, X., and Z. Cheng. 2015. Removal of diclofenac from aqueous solution with multi-walled carbon nanotubes modified by nitric acid. *Chinese Journal of Chemical Engineering* 23: 1551–1556.
87. Yang, X., R.C. Flowers, H.S. Weinberg, and P.C. Singer. 2011. Occurrence and removal of pharmaceuticals and personal care products (PPCPs) in an advanced wastewater reclamation plant. *Water Research* 45 (16) 5218–5228.
88. Kim, H., Y.S. Hwang, and V.K. Sharma. 2014. Adsorption of antibiotics and iopromide onto single-walled and multi-walled carbon nanotubes. *Chemical Engineering Journal* 255: 23–27.
89. Ding, H., X. Li, J. Wang, X. Zhang, and C. Chen. 2016. Adsorption of chlorophenols from aqueous solutions by pristine and surface functionalized single-walled carbon nanotubes. *Journal of Environmental Sciences* 43: 187–198.
90. Joseph, L., Q. Zaib, I.A. Khan, et al. 2011. Removal of bisphenol A and 17 α-ethinyl estradiol from landfill leachate using single-walled carbon nanotubes. *Water Research* 45: 4056–4068.
91. Chen, H., B. Gao, and H. Li. 2015. Removal of sulfamethoxazole and ciprofloxacin from aqueous solutions by graphene oxide. *Journal of Hazardous Materials* 282: 201–207.
92. Rostamian, R., and H. Behnejad. 2016. A comparative adsorption study of sulfamethoxazole onto graphene and graphene oxide nanosheets through equilibrium, kinetic and thermodynamic modeling. *Process Safety and Environmental Protection* 102: 20–29.
93. Dong, S., Y. Sun, J. Wu, B. Wu, A.E. Creamer, and B. Gao. 2016. Graphene oxide as filter media to remove levofloxacin and lead from aqueous solution. *Chemosphere* 150: 759–764.
94. Bele, S., V. Victoria Samanidou, and E. Deliyanni. 2016. Effect of the reduction degree of graphene oxide on the adsorption of Bisphenol A. *Chemical Engineering Research and Design* 109: 573–585.

Lignin and Chitosan-Based Materials for Dye and Metal Ion Remediation in Aqueous Systems

Thato Masilompane, Nhamo Chaukura, Ajay K. Mishra, Shivani B. Mishra and Bhekie B. Mamba

Abstract In view of dwindling fresh water sources, water pollution due to dyes and toxic metals is cause for concern. The increase in industrial activity around the world results in the emission of dyes and toxic metals into the aquatic environment and exerts pressure on water treatment plants. The removal of these contaminants is problematic because they can be available in very low concentrations, and water treatment plants are not designed to remove them effectively. A number of approaches including coagulation, precipitation, membrane filtration, and activated carbon adsorption, have been used for the remediation of contaminated water, but these methods are generally limited by high cost and poor selectivity. Lignin- and chitosan-based nanocomposites are potentially useful for these applications because they have minimal environmental footprints, are cost effective, and are compatible with a wide range of materials in composites. Laboratory scale experiments carried out to evaluate these materials have shown that the composites of these materials have remarkable dye and heavy metal (HM) removal capacities, thus making the technology accessible and potentially manageable at a large scale. Using *Web of Science*, *Scopus*, *Sciencedirect*, *Springer*, and *Google Scholar*, we evaluated literature on (1) the prevalence and environmental and health impact of pollution due to dye- and metal-laden effluents, (2) available remediation technologies, (3) the synthetic pathways for different chitosan-based nanocomposites, and (4) the potential of chitosan-based nanocomposites for dye and HM removal. There has been a gradual increase in the research of the use of lignin/chitosan-based adsorbent, showing the rapid interest and potential in the materials.

Keywords Adsorption · Biodegradable · Environment · Nanocomposites · Pollutants

---

T. Masilompane · N. Chaukura (✉) · A.K. Mishra · S.B. Mishra · B.B. Mamba
Nanotechnology and Water Sustainability Research Unit, College of Engineering, Science and Technology, University of South Africa, Johannesburg, South Africa
e-mail: nchaukura@gmail.com

A.K. Mishra
e-mail: ajaykmishra1@gmail.com

© Springer International Publishing AG 2018
S. Bhardwaj Mishra and A.K. Mishra (eds.), *Bio- and Nanosorbents from Natural Resources*, Springer Series on Polymer and Composite Materials, https://doi.org/10.1007/978-3-319-68708-7_3

1 Introduction

Although water is the largest resource, only about 1% of the world's water is available for human consumption [1, 2], the rest is contaminated with a variety of pollutants. A total of 30% of the water is used for domestic and industrial purpose while the rest is used in agriculture (Fig. 1). The ever increasing demand for industrial and consumer goods and services around the world has resulted in the emission of vast amounts of a wide range of pollutants into the aquatic environment. These include organic compounds, dyes, and heavy metals (HM) that would require removal to render the water potable or suitable for other applications. Most water treatment plants are not equipped to remove the dyes and heavy metals to satisfactory levels, consequently these pollutants persist in drinking water and cause a range of health risks to humans.

On the one hand, dyes are widely used in textile, plastic, paper, food, tannery, and cosmetic industries for colouring products. Dyes are classified according to their general structure, and textile dyes are also further classified as anionic, non-ionic, and cationic dyes [3] (Fig. 2). The major anionic dyes are the direct, acid and reactive dyes, and the most problematic ones are the brightly coloured, water soluble reactive and acid dyes.

In the textile industry, large quantities of water are used in the dyeing and printing processes, thus dye effluent is typically discharged from printing processes [4, 5]. The concentration of dyes in these effluents largely depends on the type of dye, usage rate, dye fixation degree on the substrate (i.e. textiles or metals), the extent of dye removal in the effluent treatment process, and the dilution factor in the receiving water. For instance, studies have reported the concentration of dyes to be varying between 5 and 1555 mg/L for batch process dyeing of cotton fibres with reactive dyes and 1.2–364 mg/L for batch process dyeing wool by acid dyes [6]. After treatment, the wastewater can be re-used to minimise contamination of public waterways from dyes and other auxiliary chemicals. Dye-containing effluents are undesirable because of their colour, and their degradates are toxic, carcinogenic, or mutagenic to life forms mainly because of carcinogens such as benzidine, naphthalene, and other aromatic compounds [7, 8]. The effluent imparts an undesirable colour which limits the penetration of light and hence prevents photosynthetic activity in aquatic plants and ultimately affects aquatic life [9]. Moreover, many

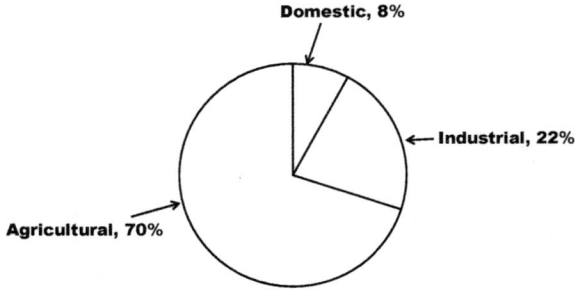

Fig. 1 Fresh water consumption in various activities

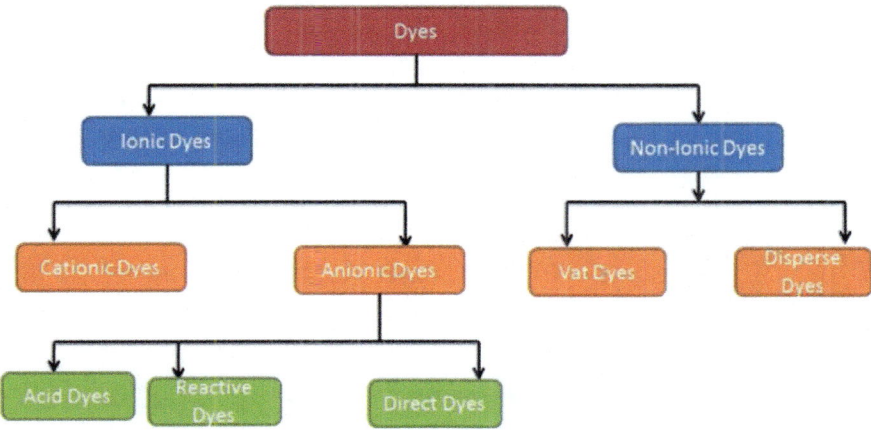

Fig. 2 Examples of dyes and their classification [3]. Reprinted from Tan et al. [3], Copyright (2015), with permission from Elsevier

dyes are not biodegradable and persist in the environment creating serious water quality and public health problems. If allowed to flow in the fields, the effluents clog the pores in the soil, altering soil texture, and resulting in decreased productivity. In drains and rivers, it affects the quality of drinking water in aquifers and hand pumps making it unfit for human consumption. Besides, such polluted water can also be a breeding ground for bacteria and viruses. Conventional biological treatment processes are not very effective in treating dyes wastewater [10]. These methods include oxidation, coagulation and flocculation, biological treatment, and membrane filtration [3, 8, 11–14] (Fig. 3).

The major disadvantages of these methods include the production of toxic sludge, high operational cost, technical complexity, ineffective colour reduction, and sensitivity to variation in influent quantity and quality [15, 16]. In addition, a single conventional treatment is unable to remove certain forms of colour. Adsorption is an attractive and effective alternative technique for the removal of dyes from contaminated water for a number of reasons: (1) it has the ability to remove a wide range of pollutants [8, 17], (2) it is generally cheaper, especially when low-cost adsorbents are used [2, 18], and (3) it is relatively easier to regenerate spent adsorbent [19]. Adsorption can be achieved by use of natural and synthetic adsorbents, which include activated carbon [3], alumina [18], silica gel [12], industrial waste products [9], metal organic frameworks [17], zeolites [20], clay minerals [14, 19], and agro wastes [4, 5, 21]. Agricultural and industrial wastes have the advantage of being readily and cheaply available, and in most cases, they are not easily disposed and thus cause an environmental and public health risk [22]. Using these wastes for removal of pollutants from contaminated water systems would help in their disposal. A major advantage of using the sorption technique is being able to use waste to clean-up another waste hence resulting in environmental sustainability [23].

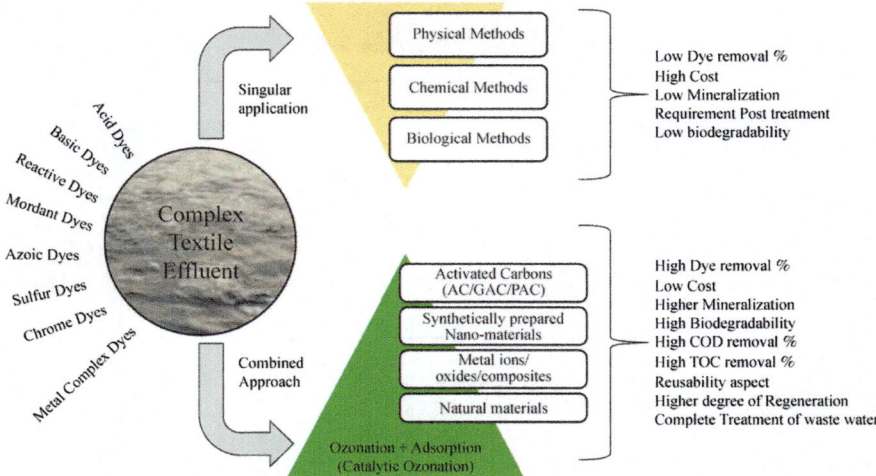

Fig. 3 Dye removal technologies. Khamparia and Jaspal [84] with permission of Springer

On the other hand, water pollution due to heavy metals (HM) is extensively documented (e.g. [24–26]). Despite lack of consensus on the definition of HM, it is generally accepted that these are metal elements with a density exceeding 5 g/cm^3 [26, 27]. Sources of HM include anthropogenic activities such as metalliferous mining operations, metal plating, metallurgical facilities, coal fly ash from power stations, electronic and electrical wastes, pigments and paint industries, pesticide industries, and sewage sludge [25, 28, 29]. Although a few HMs like Co, Cr, Cu, Mn, Mo, and Zn are necessary in small concentrations for the growth of plants and animals, HMs that are bioavailable in excessive levels have potential to cause toxicity to aquatic organisms, and up the food chain to humans through agricultural produce and drinking water [26, 30–32]. HM pollution adversely affects biota, humans, and disrupts the ecosystem. Furthermore, because they are not biodegradable, most heavy metals bioaccumulate in the environment and in human tissue, through food chain transfer, resulting in an assortment of environmental and health problems [24, 26]. The HMs of major concern pertaining to human health and environmental impacts are As, Cd, Hg, Pb, Tl, and U. For example, in its organic form, U mainly causes neurotoxicity and teratogenicity, while acute Zn exposure has been associated with lethargy, tremors, and growth retardation in mammals [26, 33], and exposure to Cd and Pb can lead to renal damage, osteoporosis, and bone defects [28, 29]. Extensive human epidemiological data showed that inhaled As leads to lung cancer and ingested As leads to skin cancer [31]. Research has shown that, because children have a permeable blood-brain barrier, they are more prone to Pb poisoning, which can result in defects to the central nervous system [26]. HMs rarely occur singly in the environment, and as such, their coexistence can cause synergistic or additive toxicity to organisms. For instance, excessive uptake of Pb, Hg, and Al can reduce the metabolism of essential nutrients

Table 1 Impacts of heavy metals on humans

Heavy metal	Affected organs	Health effect
Cd	Kidneys	Chronic exposure to Cd can increase the risk of renal cancer
Pb	Skin	Lead concentrations are associated with disease severity and the health of children with skin diseases
As	Lungs, bladder, kidney, liver, skin	Prolonged exposure to As from drinking water causes cancers in lung, urinary bladder, kidney, liver, and skin
Cu, Zn	Pancreas	Low serum Zn levels and elevated Cu concentrations are connected to chronic pancreatitis
Cr	Pancreas	Cr concentrations that have strong association with pancreatic cancer

Adapted from Wijayawardena et al. [26], Copyright (2016), with permission from Elsevier

like Ca, while Cd competes with Zn for specific binding sites on metallothionein that is required for Zn transport [26]. A more extensive presentation of HM toxicity in humans is in Table 1.

Commonly used techniques for HM removal from aqueous systems include precipitation, coagulation, membrane filtration, and ion exchange on resins [28, 34]. Other potential but less common methods include capacitative deionization [35], electrokinetic remediation [36], ultrafiltration [37], reverse osmosis [24], and use of zero valent iron (ZVI) [38]. Whereas adsorption is the most attractive because of low cost and high efficiency of removal from dilute solution [33], ion exchange is efficient when large volumes of wastewater containing trace quantities are treated. Moreover, ion exchange enjoys the advantage of simplicity and is suitable for operation under a wide range of reaction conditions. Materials that include biopolymers like chitosan and lignin have been widely explored for their ability to remove various pollutants from aqueous systems [39, 40]. The benefits of these systems include (1) both chitosan and lignin being biopolymers are biodegradable and therefore environment-friendly, (2) they are abundant in nature and can therefore be obtained cost-effectively, and (3) incorporating nanomaterials can significantly improve the physico-chemical properties of the materials, thus making them more suitable as potential adsorbents for a wide range of pollutants.

2 Chitosan and Lignin

There has been a rapid increase in research activity on chitosan- and lignin-based water remediation approaches, and significant findings have been reported. On the one hand, chitosan is one of the most abundant naturally occurring biopolymer resulting from the alkaline *N*-deacetylation of the chitin constituent of the exoskeletons of shellfish and crustaceans, producing a linear cationic polymer of alpha (1–4) -linked 2-amino-2-deoxy-beta-*D*-glucopyranose [41, 42] (Fig. 4).

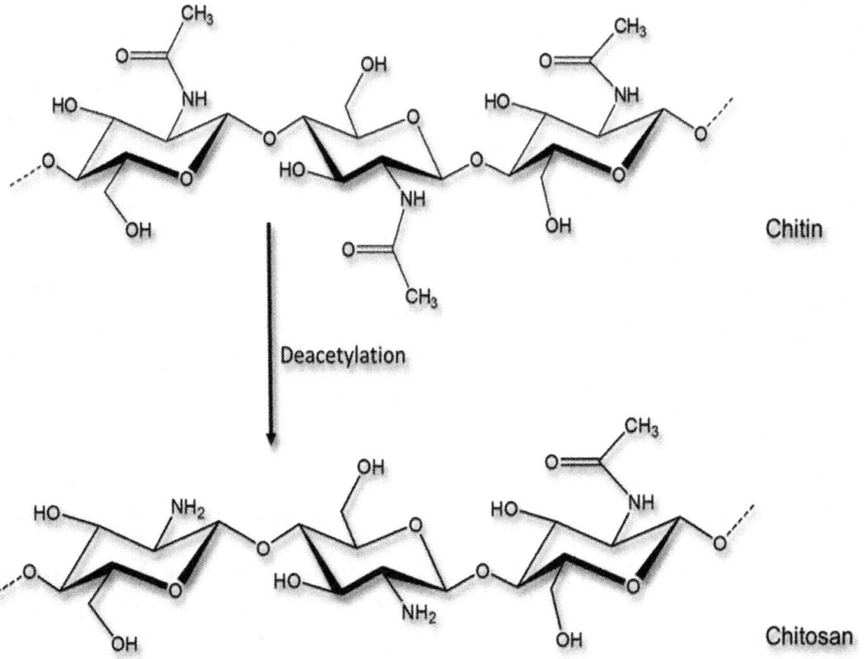

Fig. 4 Preparation of chitosan from chitin by deacetylation

Chitin is a highly crystalline biopolymer composed of beta (1–4) -linked 2-acetamide-2-dexy-beta-*D*-glucose (*N*-acetyl glucosamine) motifs, giving chitosan useful functional groups such as OH and –NH_2 groups that can be modified to various moieties [43]. Consequently, the chemistry of chitosan is considerably more robust than that of cellulose due to the –NH_2 amino on the chitosan [43]. Its key characteristics include the molecular weight, degree of deacetylation, and crystallinity [44].

Although chitin does not occur in organisms that produce cellulose, it is considered as cellulose derivative as it is structurally identical to cellulose. It therefore enjoys similar properties such as biodegradability, biocompatibility, non-toxicity, adsorption properties, and hydrophilicity. Owing to their physico-chemical properties, chitinous materials are versatile biomaterials due to their diverse bioactivities, biocompatibility, biodegradability, non-toxicity, and low allergenicity [44]. Moreover, chitosan can easily be fabricated into films and membranes that can be used for water purification by filtration [45].

The presence of –NH_2, –OH, and –O– groups on the chitosan enables it to adsorb metal ions and dyes through a number of mechanisms that include ion exchange, co-ordination, hydrogen bonding, and Van der Waal's forces, electrostatic forces, and weak intermolecular forces [41]. In acidic medium, the amino group on the chitosan can be protonated, allowing it to strongly adsorb anions by electrostatic attraction. Albadarin et al. [46] investigated methylene blue adsorption onto activated lignin-chitosan extrudates (ALiCE) and proposed a mechanism that involves electrostatic attraction between a quaternary nitrogen on the methylene blue and an oxygen anion on ALiCE (Fig. 5). Chitosan thus acts as a chelating agent, binding metal ions onto the heteroatom bearing groups and removing them from solution. Studies have shown that chitosan can be an effective biosorbent for removal of metal anions like Cr^{6+} [47] and anionic dyes such as acid, reactive, and direct dyes [43].

On the other hand, lignin is a biodegradable, non-toxic, bioactive natural polymer derived from lignocellulosic biomass, which constitutes about 50% of all biomass ([48, 49]. It is abundant in the cell walls of terrestrial plants, typically constituting 16–33% of plant biomass. Lignin is a random aromatic three-dimensional network copolymer of phenyl-propane units (coniferyl, sinapyl, and p-coumaryl alcohol) and serves as the matrix or binding agent for cellulose and hemicellulose [50]. The main functional groups in lignin are $-OH$, $-OCH_3$, $-CO$, $-O-$, and $-COOH$ in various proportions that depend on the origin and extraction processes [51]. The linkage structure of lignin depends on its botanical origin and on the extraction method [52]. For example, wood lignin is connected to plant cell wall polysaccharides through benzyl ether, benzyl ester, and phenyl glycoside linkages [53], whereas in annual plants, ester linkages are more abundant [54] (Fig. 6). The carboxyl and phenol groups contribute to the pH-dependent sorption mechanism, and phenolic groups have a higher affinity towards metal ions [51]. The monolayer adsorption capacity is dependent on the ionic charge and hydrodynamic radius of the metal pollutant, so that, while the maximum adsorption capacity for Cr^{6+} is modest, that for Cr^{3-} is fairly high [55], and the maximum adsorption

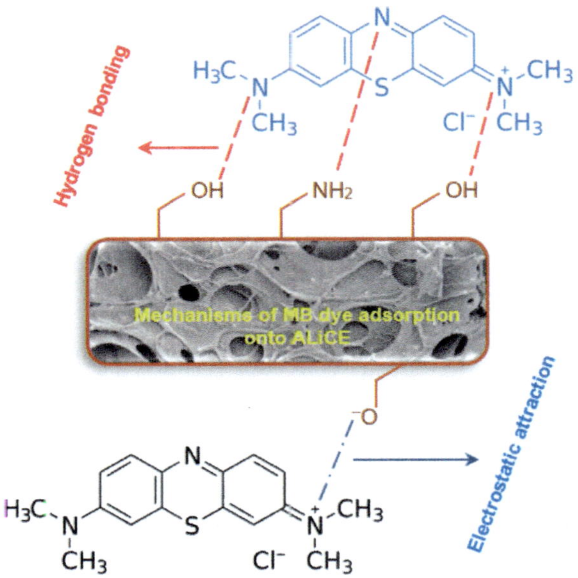

Fig. 5 Proposed methylene blue adsorption mechanism on a chitosan-derived adsorbent, ALiCE. Reprinted from Albadarin et al. [46], Copyright (2017), with permission from Elsevier

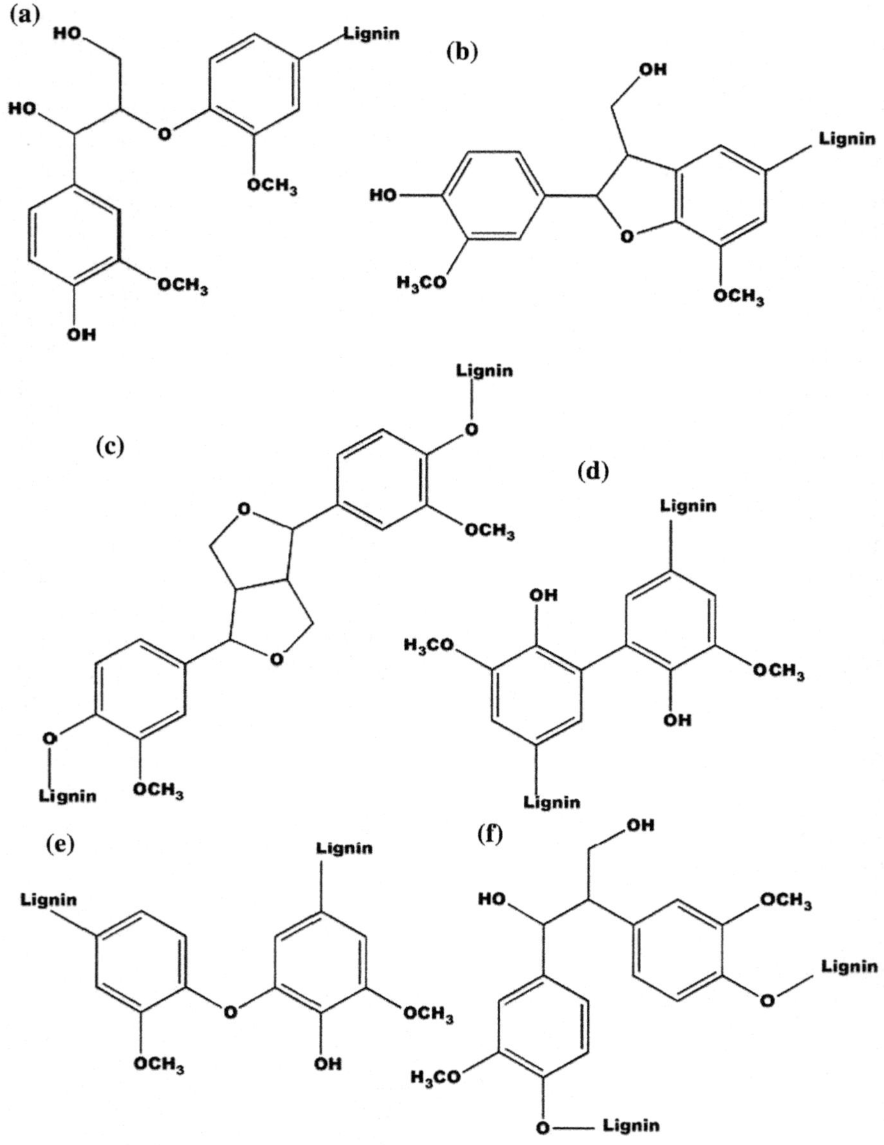

Fig. 6 Common interunits found in the lignin structure, namely (**a**) β-aryl-ether (β-O-4 linkage), (**b**) phenylcoumaran (β-5 linkage), (**c**) resinol (β-β' linkage), (**d**) biphenyl (5-5' linkage), (**e**) diaryl ether (5-O-4 linkage) and f) diphenyl ether (β-1' linkage)

capacity for divalent cations follows the trend Pb > Cu > Cd > Zn > Ni [56, 57]. For both lignin and chitosan-based adsorbents, the adsorption kinetics for both HMs and dyes invariably followed pseudo-second-order kinetics (Table 1).

Besides, lignin has also been reported to enhance the mechanical properties of plastic and edible films, mainly their barrier properties [49]. Thus, lignin with a surface area of 180 m^2/g and many functional groups which can serve as adsorption sites for the uptake of both organic and inorganic contaminants from aqueous systems.

The sterically exposed oxygen groups are expected to interact with heavy metals and cationic dyes, while the phenolic groups can interact with benzenes, chlorophenol, and azo dyes [58]. Because of the presence of a π-electron system in the polymer backbone, there is also possibility for π-π interactions between lignin and pollutants, for example dyes [3]. These functional groups make lignin an attractive adsorbent for water treatment applications. Moreover, lignin can be converted to activated carbon or biochar through physical or chemical means [59]. Chemical activation at elevated temperatures produces higher product yields. Activated carbons have been extensively studied for their application in soil conditioning, wastewater remediation, and contaminated soil remediation [60]. Although lignin offers the benefits of abundance, low cost, good chemical and thermal stability, and adequate O-carrying groups, the major constraint is the depressed surface area, which results in low response rates and decreased adsorption capacities that limit its application in HM or dye removal [61].

Although chitosan/lignin-based contaminated water remediation approaches have been extensively studied [18, 29, 62], there is limited literature on the wastewater remediation applications of chitosan/lignin composites with other materials. There is thus need to explore the use of composites derived from lignin in order to improve the properties of these materials.

2.1 Chemical Modification

Lignin can easily be chemically modified by introducing new active sites, or by functionalising of the hydroxyl functional group to introduce desirable moieties for the removal of specific pollutants [62] (Fig. 7). These modifications enhance the chemical reactivity and the physical properties of lignin. Such physical properties include improved solubility of the otherwise recalcitrant molecule that is difficult to process [63]. For example alkyl and alkoxy groups can be introduced through nucleophilic aromatic substitution of the phenolic OH. Esterification, phenolation, and etherification are also possible modifications that can be exploited. Other functionalities that can also be introduced include amines, nitrates, and the sulphonic groups. In its native or functionalised state, lignin has been used for the effective adsorption of HM like chromium, cobalt, copper, cadmium, lead, nickel, and zinc [51].

Chitosan, on the other hand, undergoes acrylation to produce materials that have adhesive properties [64]. Moreno-Vazquez et al. [65] modified chitosan via a free radical reaction and produced a material with antioxidant and antibacterial properties. The commonly used chemical modification of chitosan is the

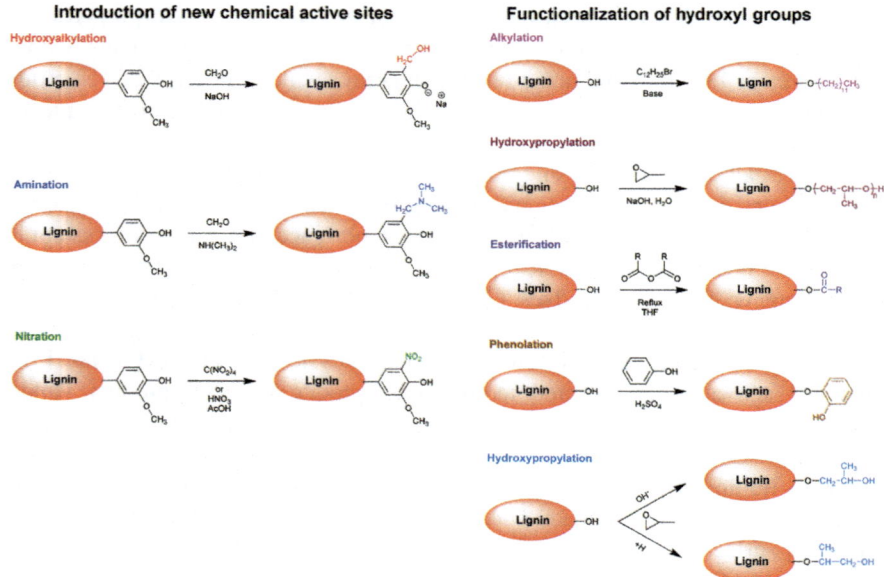

Fig. 7 Chemical modification of lignin. Reproduced from Kai et al. [62] with permission from the Royal Society of Chemistry

substitution of –NH$_2$ groups with motifs that confer it with desired properties [66]. To this effect, carbohydrates have been introduced to improve the solubility of chitosan. It could also be possible to introduce rigid molecular units onto the polymer backbone in order to enhance porosity and surface area, which properties are desirable in pollutant adsorption (Fig. 8).

2.2 Composites

Owing to their unique physico-chemical properties, polymer nanocomposites comprising metal-based nanoreinforcement and a biopolymer framework have received significant research attention. The properties of chitosan and lignin can be remarkably improved by incorporating other materials in blends or composites. Because lignin has phenyl and hydroxyphenyl groups, it shows catalytic properties, a desirable characteristic in chemically modified composites [67]. The development nanocomposites enhance the surface area, surface functionality, morphology, and thermal stability of the materials and improve the adsorption capacities [41]. Examples of blends include lignin–chitosan [68], starch–lignin blends [69], and lignin poly(acrylic acid) copolymers [70], while composites range from metal oxide composites such as TiO$_2$-lignin, TiO$_2$-chitosan [41], and magnetic composites [40].

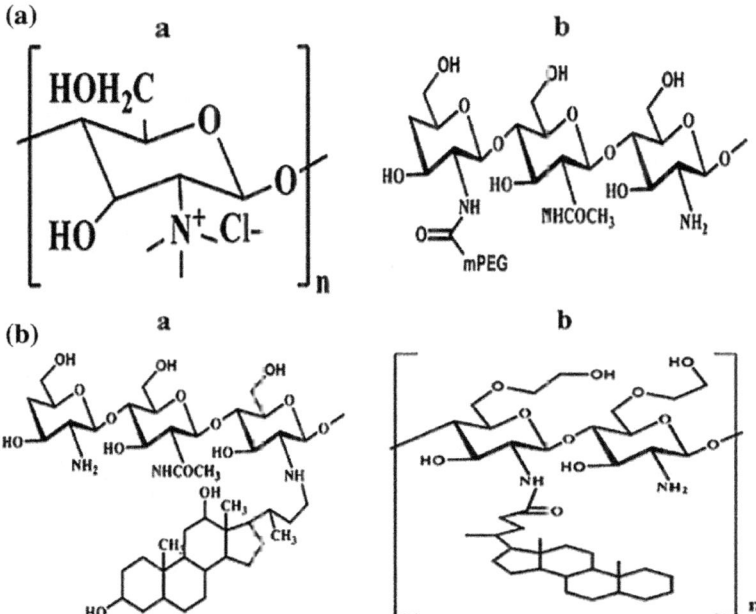

Fig. 8 Chitosan derivatives with hydrophilic group (A) or hydrophobic group (B). A (a) trimethylated chitosan (b) pegylated chitosan. B (a) deoxycolic acid modified chitosan, (b) 5β-cholanic acid-modified chitosan. Choi et al. [85]. Reprinted from Choi et al. [85], Copyright (2016), with permission from Elsevier

2.2.1 Synthesis and Properties of Composites

Various methods such as hydrothermal, direct oxidation, chemical vapour deposition, sonochemical, and sol-gel methods have been employed in the synthesis of TiO_2-lignin and TiO_2-chitosan nanoparticles [71]. However, the most commonly used technique for preparing chitosan nanoparticles is the use of an anionic compound such as alginate, tripolyphosphate, carrageenan, or polyelectrolyte [71]. The sol-gel method remains the most used method due to advantages such as control over morphology and structure of the resultant nanoparticles. The sol-gel technique is used for fabricating advanced materials, including ceramics and organic–inorganic hybrids. In general, it is a process that involves the transition of a solution system from liquid to sol into solid gel phase. Advantages of the sol-gel method include use of relatively low temperatures, stoichiometry control, creation of very fine powders, and production of high purity compositions that are otherwise not possible by solid-state fusion [72].

The properties, and hence applications of the composites depend on both the method of synthesis and the constituents used. Lignin nanoparticles have been synthesised through chemical, physical, and mechanical techniques that use Kraft, acetylated, phosphorylated lignins [73]. For instance, chitosan/clay nanocomposites were synthesised using magnetic nanoparticles and clay encapsulated in cross-linked chitosan beads and used for the removal of methylene blue dye [74]. This is because clays are layered structures with water molecules intercalated in-between, so that the interlayer spaces can expand and adsorb fugitive molecules [75]. These spaces have well-defined geometry that confers specificity towards particular pollutants depending on molecular size. Clays can themselves be activated by introducing specific functional groups or by making composites with other materials such as chitosan, thus enhancing the specificity. In a separate study, Zeng et al. [76] synthesised a three-component composite from chitosan, polyaluminium chloride (PAC), and silicate and produced a 7–34% cheaper flocculant with between 61.2–85.5% Al^{3+} removal enhancement compared to PAC. Zhou et al. [77] studied the preparation and application of magnetic chitosan–lignin fibre (mCS/LCF) composites for the removal of Acid Red 18 (AR18) dye and reported removal efficiencies of up to 99.4%. Using flash precipitation approach, tunable surface properties were obtained on composites derived from Kraft lignin and Organosolv lignin with poly(diallyldimethylammonium chloride) coating, which had nanoparticles in the size range of 45–250 nm [78]. Nitayaphat and Jintakosol [79] investigated the adsorption of Ag^+ using chitosan/bamboo composite beads and reported a maximum adsorption capacity of 52.91 mg/g. Wang and Wang [80] synthesised a chitosan-poly(vinyl alcohol)/attapulgite composite which removed Cu^{2+} with a maximum adsorption capacity of 35.79 mg/g. In another study, lignin microspheres were prepared via an inverse suspension copolymerization technique and demonstrated effective Pb^{2+} removal in aqueous systems [18]. A remarkably high Cd^{2+} maximum adsorption capacity of 833.3 mg/g was obtained from chitosan-graft-γ-cyclodextrin (Ch-g-γ-CD) composites at an optimum pH of 8.5 [81]. The adsorption capacity is also a function of the functional groups present on the adsorbent surface. For instance, the maximum adsorption capacity of lignin for Ni^{2+} removal was increased by 36% by modifying the lignin with phenol to make lingo-phenol [82]. In a related study, there was a 68% increase in Pb^{2+} removal capacity for TiO_2/lignin after incorporating SiO_2 into the polymer structure to make a three-component TiO_2–SiO_2/lignin composite [83]. The reason for this was the increase in the presence of more oxygen-carrying groups and larger pore volumes on the latter composite. Due to the presence of a high density of oxygen-carrying functional groups, chitosan–lignin composites showed very high Acid Red 18 removal capacity (1184 mg/g) [77]. This, however, is in contrast to the relatively low methylene blue removal capacity for activated chitosan–lignin composites (36.25 mg/g) [46], due to the differences in the chemistry of the two dyes. A summary of the properties of selected composites is presented in Table 2. Overall, the incorporation of lignin and chitosan into nanocomposites significantly improves the surface functional groups and hence the HM and dye removal capacities.

Table 2 Properties of chitosan- and lignin-based adsorbents

Adsorbate Heavy metals	Adsorbent	Adsorption capacity (mg/g)	Remarks	Source
Pb^{2+}	TiO_2/lignin, TiO_2–SiO_2/lignin	35.70, 59.93	Maximum volume contribution 531–1720 nm particles, 955 nm–25 0%, 106–295, 1480–5560 nm particles, 2670 nm–13.5%, pseudo-second-order kinetics	[83]
Ag^+	CTS/bamboo charcoal	52.91	S_{BET}: 34.34, highest adsorption efficiency when the bamboo charcoal at 50% (w/w) charcoal	[79]
Cu(ll)	CTS-PVA/APT	35.79	pH: 2.0–6.5, pseudo-second-order kinetics	[80]
Pb	Lignin microspheres	33.9	Diameter: 248 μm, S_{BET}: 9.6, abundant –NH_2 functional groups	[61]
Ni(ll), Cd (II)	Silica/lignin	77.11, 84.66	S_{BET}: 223, pseudo-second-order kinetics	[57]
Cr(lll)	Lignin	17.97	Lignin isolated from paper and pulp black liquor. SBET: 21.7	[55]
Cr(Vl)	CTS	7.94	Optimum pH: 3, pseudo-second-order kinetics	[47]
Ni(ll)	Modified lignin	16.94	Lignin extracted from *T. diversifolia* biomass and modified with phenol to make ligno-phenol, pseudo-second-order kinetics	[82]
Ni(ll)	Lignin	12.48	Lignin extracted from *T.diversifolia* biomass, pseudo-second-order kinetics	[82]
Cd(ll)	CTS-g-y-cyclodextrin	833.3	Optimum pH: 8.5, 238% grafting	[81]
Dyes				
RBBR	CTS-alkaline lignin	111.1	Chitosan–alkali lignin (50:50) composite exhibited maximum percentage removal of anthraquinonic dye, S_{BET}: 2.4403	[86]
Acid Red 18	CTS-lignocellulose fibre	1184	Magnetized chitosan-coated lignocellulose fibre, pH: 3, pseudo-second-order kinetics	[58]
Amino black 10B	CTS/bentonite	990.1	Prepared by dispersing bentonite in quaternized chitosan by membrane-forming method followed by cross-linking	[19]

(continued)

Table 2 (continued)

Adsorbate Heavy metals	Adsorbent	Adsorption capacity (mg/g)	Remarks	Source
Methyl blue	Activated lignin–chitosan extruded (ALiCE)	36.25	Activated lignin–chitosan pellets	[46]
Reactive Black 5	CTS	696.99	Degree of deacetylation: 85%	[87]
Direct red 81	CTS beads	2383	Chemically cross-linked chitosan	[1]

S_{BET} is BET surface area (m^2/g), and CTS is chitosan

3 Summary

Chitosan- and lignin-based contaminated water remediation approaches have been widely studied. The benefits of these systems include (1) both chitosan and lignin being biopolymers are biodegradable and therefore environment-friendly, (2) they are abundant in nature and can therefore be obtained cost-effectively, and (3) incorporating nanomaterials can significantly improve the physico-chemical properties of the materials, thus making them more suitable as potential adsorbents for a wide range of pollutants. Both chitosan and lignin can be chemically modified into new materials with various functional groups and properties, making them useful for a whole host of applications. Although there is considerable research on lignin and chitosan separately, there is limited literature on the applications of chitosan/lignin composites. The existing research on the chitosan–lignin composites demonstrates that the materials have high affinity for both HMs and dyes. Chitosan and lignin materials are compatible with nanomaterials, and, as such, have been incorporated into nanocomposites for the removal of pollutants from polluted water. Both these materials introduce oxygen-carrying moieties that enhance the adsorption of cationic species.

4 Future Scenario

In view of the limited research on the application of lignin/chitosan composites, there is need to investigate (1) the properties of lignin/chitosan-based materials, and (2) their applications for wastewater remediation targeting a range of metallic and organic pollutants. These materials form the platform for the next generation of water remediation technologies that have minimal environmental footprint.

References

1. Gupta, V.K., P.J.M. Carrott. M.M.L. Carrott, and Suhas. 2009. Low-cost adsorbents: Growing approach to wastewater treatment—A review. *Critical Reviews in Environmental Science and Technology* 39: 783–842.
2. Adeleye, A.S., J.R. Conway, K. Garner, Y. Huang, Y. Su, and A.A. Keller. 2016. Engineered nanomaterials for water treatment and remediation: Costs, benefits, and applicability. *Chemical Engineering Journal* 286: 640–662.
3. Tan, K.B., M. Vakili, B.A. Horri, P.E. Poh, A.Z. Abdullah, and B. Salamatinia. 2015. Adsorption of dyes by nanomaterials: Recent developments and adsorption mechanisms. *Separation and Purification Technology* 150: 229–242.
4. Sewu, D.D., P. Boakye, and S.H. Woo. 2017. Highly efficient adsorption of cationic dye by biochar produced with Korean cabbage waste. *Bioresource Technology* 224: 206–213.
5. Shen, K., and M.A. Gondal. 2017. Removal of hazardous Rhodamine dye from water by adsorption onto exhausted coffee ground. *Journal of Saudi Chemical Society* 21: S120–S127.
6. Zaharia, C. and D. Suteu. 2012. Textile organic dyes characteristics, polluting effects, and separation/elimination procedures from industrial effluents. A critical overview. In *Organic pollutants—Ten years after the Stockholm convention. Environmental and analytical update*, ed. T. Puzyn and A. Mostrag-Szlichtyng, 55–86. Rijeka: Intech Publisher Inc.
7. Meili, L., T.S. da Silva, D.C. Henrique, J.I. Soletti, S.H.V. de Carvalho, E.J.S. Fonseca, A.R.F. de Almeida, and G.L. Dotto. 2016. Ouricuri (*Syagrus coronata*) fiber: A novel biosorbent to remove methylene blue from aqueous solutions. *Water Science and Technology*. doi:10.2166/wst.2016.495.
8. Blanco, S.P.D.M., F.B. Scheufele, A.N. Módenes, F.R. Espinoza-Quiñones, P. Marin, A.D. Kroumov, and C.E. Borba. 2017. Kinetic, equilibrium and thermodynamic phenomenological modeling of reactive dye adsorption onto polymeric adsorbent. *Chemical Engineering Journal* 307: 466–475.
9. Belhaine, A., M.R. Ghezzar, F. Abdelmalek, K. Tayebi, A. Ghomari, and Addou A. Ahmed. 2016. Removal of methylene blue dye from water by a spent bleaching earth biosorbent. *Water Science and Technology*. doi:10.2166/wst.2016.407.
10. Kyzas, G.Z., E.A. Deliyanni, and N.K. Lazaridis. 2014. Magnetic modification of microporous carbon for dye adsorption. *Journal of Colloid and Interface Science* 430: 166–173.
11. Abdel-Khalek, M.A., M.K.A. Rahman, and A.A. Francis. 2017. Exploring the adsorption behavior of cationic and anionic dyes on industrial waste shells of egg. *Journal of Environmental Chemical Engineering* 5: 319–327.
12. Han, H., W. Wei, Z. Jiang, J. Lu, J. Zhu, and J. Xie. 2016. Removal of cationic dyes from aqueous solution by adsorption onto hydrophobic/hydrophilic silica aerogel. *Colloids and Surfaces A: Physicochem. Engineering. Aspects* 509: 539–549.
13. Bhattacharyya, A., D. Mondal, I. Roy, G. Sarkar, N.R. Saha, D. Rana, T.K. Ghosh, D. Mandal, M. Chakraborty, and D. Chattopadhyay. 2017. Studies of the kinetics and mechanism of the removal process of proflavine dye through adsorption by graphene oxide. *Journal of Molecular Liquids* 230: 696–704.
14. Ngulube, T., J.R. Gumbo, V. Masindi, and A. Maity. 2017 An update on synthetic dyes adsorption onto clay based minerals: A state-of-art review. *Journal of Environmental Management* 191: 35–57.
15. Lijo, L., S. Malamis, S. Gonzalez-García, F. Fatone, M.T. Moreira, and E. Katsou. 2017. Technical and environmental evaluation of an integrated scheme for the co-treatment of wastewater and domestic organic waste in small communities. *Water Research* 109: 173–185.
16. Raman, C.D., and S. Kanmani. 2016. Textile dye degradation using nano zero valent iron: A review. *Journal of Environmental Management* 177: 341–355.

17. Luo, X., S. Fu, Y. Du, J. Guo, and B. Li. 2017. Adsorption of methylene blue and malachite green from aqueous solution by sulfonic acid group modified MIL-101. *Microporous and Mesoporous Materials* 237: 268–274.
18. Wasti, A., and M.A. Awan. 2016. Adsorption of textile dye onto modified immobilized activated alumina. *Journal of the Association of Arab Universities for Basic and Applied Sciences* 20: 26–31.
19. Zhang, L., P. Hu, J. Wang, and R. Huang. 2016. Crosslinked quaternized chitosan/bentonite composite for the removal of Amino black 10B from aqueous solutions. *International Journal of Biological Macromolecules* 93: 217–225.
20. Aysan, H., S. Edebali, C. Ozdemir, M.C. Karakaya, and N. Karakaya. 2016. Use of chabazite, a naturally abundant zeolite, for the investigation of the adsorption kinetics and mechanism of methylene blue dye. *Microporous and Mesoporous Materials* 235: 78–86.
21. Kumar, A., and H.M. Jena. 2016. Removal of methylene blue and phenol onto prepared activated carbon from Fox nutshell by chemical activation in batch and fixed-bed column. *Journal of Cleaner Production* 137: 1246–1259.
22. Mohan, D., K.P. Singh, and V.K. Singh. 2008. Wastewater treatment using low cost activated carbons derived from agricultural byproducts—A case study. *Journal of Hazardous Materials* 152: 1045–1053.
23. Elaigwu, S.E., and G.M. Greenway. 2016. Microwave-assisted and conventional hydrothermal carbonization of lignocellulosic waste material: Comparison of the chemical and structural properties of the hydrochars. *Journal of Analytical and Applied Pyrolysis* 118: 1–8.
24. Kalaivani, S.S., A. Muthukrishnaraj, S. Sivanesan, and L. Ravikumar. 2016. Novel hyperbranched polyurethane resins for theremoval of heavy metal ions from aqueous solution. *Process Safety and Environmental Protection* 104: 11–23.
25. Kang, C., and J. So. 2016. Heavy metal and antibiotic resistance of ureolytic bacteria and their immobilization of heavy metals. *Ecological Engineering* 97: 304–312.
26. Wijayawardena M.A.A., Megharaj M., Naidu R. (2016) Exposure, toxicity, health impacts, and bioavailability of heavy metal mixtures. In *Advances in Agronomy*, vol. 138. http://dx.doi.org/10.1016/bs.agron.2016.03.002.
27. Mustafa, G., and S. Komatsu. 2016. Toxicity of heavy metals and metal-containing nanoparticles on plants. *Biochimica et Biophysica Acta* 1864: 932–944.
28. Bunhu, T., L. Tichagwa, and N. Chaukura. 2016. Competitive sorption of Cd^{2+} and Pb^{2+} from a binary aqueous solution by poly (methyl methacrylate)-grafted montmorillonite clay nanocomposite. *Appl Water Sci.* doi:10.1007/s13201-016-0404-5.
29. Zhang, J., L. Li, Y. Li, and C. Yang. 2017. Microwave-assisted synthesis of hierarchical mesoporous nano-TiO_2/cellulose composites for rapid adsorption of Pb^{2+}. *Chemical Engineering Journal* 313: 1132–1141.
30. Wu, Q., H. Zhou, N.F.Y. Tam, Y. Tian, Y. Tan, S. Zhou, Q. Li, Y. Chen, and J.Y.S. Leung. 2016. Contamination, toxicity and speciation of heavy metals in an industrialized urban river: Implications for the dispersal of heavy metals. *Marine Pollution Bulletin* 104: 153–161.
31. Gupta, P., and B. Diwan. 2017. Bacterial Exopolysaccharide mediated heavy metal removal: A Review on biosynthesis, mechanism and remediation strategies. *Biotechnology Reports* 13: 58–71.
32. Korashy, H.M., I.M. Attafi, K.S. Famulski, S.A. Bakheet, M.M. Hafez, A.M.S. Alsaad, and A.R.M. Al-Ghadeer. 2017. Gene expression profiling to identify the toxicities and potentially relevant human disease outcomes associated with environmental heavy metal exposure. *Environmental Pollution* 221: 64–74.
33. Cai, Y., C. Li, D. Wu, W. Wang, F. Tan, X. Wang, P.K. Wong, and X. Qiao. 2017. Highly active MgO nanoparticles for simultaneous bacterial inactivation and heavy metal removal from aqueous solution. *Chemical Engineering Journal* 312: 158–166.
34. Kiran, M.G., K. Pakshirajan, and G. Das. 2017. Heavy metal removal from multicomponent system by sulfate reducing bacteria: Mechanism and cell surface characterization. *Journal of Hazardous Materials* 324: 62–70.

35. Huang, J., F. Yuan, G. Zeng, X. Li, Y. Gu, L. Shi, W. Liu, and Y. Shi. 2017. Influence of pH on heavy metal speciation and removal from wastewater using micellar-enhanced ultra filtration. *Chemosphere* 173: 199–206.
36. Xu, Y., C. Zhang, M. Zhao, H. Rong, K. Zhang, and Q. Chen. 2017. Comparison of bioleaching and electrokinetic remediation processes for removal of heavy metals from wastewater treatment sludge. *Chemosphere* 168: 1152–1157.
37. Robinson, T. 2017. Removal of toxic metals during biological treatment of landfill leachates. *Waste Management*. doi:10.1016/j.wasman.2016.12.032.
38. Dong, H., Y. Chen, G. Sheng, J. Li, J. Cao, Z. Li, and Y. Li. 2016. The roles of a pillared bentonite on enhancing Se(VI) removal by ZVI and the influence of co-existing solutes in groundwater. *Journal of Hazardous Materials* 304: 306–312.
39. Ngah, W.S.W., L.C. Teong, and M.A.K.M. Hanafiah. 2011. Adsorption of dyes and heavy metal ions by chitosan composites: A review. *Carbohydrate Polymers* 83: 1446–1456.
40. Kolodyńska, D., M. Gęca, I.V. Pylypchuk, and Z. Hubicki. 2016. Development of New Effective Sorbents Based on Nanomagnetite. *Nanoscale Research Letters* 11: 152–162.
41. Tran, V.S., H.H. Ngo, W. Guo, J. Zhang, S. Liang, C. Ton-That, and X. Zhang. 2015. Typical low cost biosorbents for adsorptive removal of specific organic pollutants from water. *Bioresource Technology* 182: 353–363.
42. Ghaee, A., J. Nourmohammadi, and P. Danesh. 2017. Novel chitosan-sulfonated chitosan-polycaprolactone-calcium phosphate nanocomposite scaffold. *Carbohydrate Polymers* 157: 695–703.
43. Srinivasan, A., and T. Viraraghavan. 2010. Oil removal from water using biomaterials. *Bioresource Technology* 101: 6594–6600.
44. Renault, F., B. Sancey, P.M. Badot, and G. Crini. 2009. Chitosan for coagulation/flocculation processes—An eco-friendly approach. *European Polymer Journal* 45: 1337–1348.
45. Ammar, N.S., H. Elhaes, H.S. Ibrahim, W. El hotaby, and M.A. Ibrahim. 2014. A novel structure for removal of pollutants from wastewater. *Spectrochimica Acta Part A: Molecular and Biomolecular Spectroscopy* 121: 216–223.
46. Albadarin, A.B., M.N. Collins, M. Naushad, S. Shirazian, G. Walker, and C. Mangwandi. 2017. Activated lignin-chitosan extruded blends for efficient adsorption of methylene blue. *Chemical Engineering Journal* 307: 264–272.
47. Aydin, Y.A., and N.D. Aksoy. 2009. Adsorption of chromium on chitosan: Optimization, kinetics and Thermodynamics. *Chemical Engineering Journal* 151: 188–194.
48. Gupta, V.K., and Suhas. 2009. Application of low-cost adsorbents for dye removal—A review. *Journal of Environmental Management* 90: 2313–2342.
49. Gong, W., Z. Ran, F. Ye, and G. Zhao. 2017. Lignin from bamboo shoot shells as an activator and novel immobilizing support for α-amylase. *Food Chemistry* 228: 455–462.
50. Santos, P.S.B., X. Erdocia, D.A. Gatto, and J. Labid. 2014. Characterisation of Kraft lignin separated by gradient acid precipitation. *Industrial Crops and Products* 55: 149–154.
51. Duval, A., and M. Lawoko. 2014. A review on lignin-based polymeric, micro- and nano-structured Materials. *Reactive and Functional Polymers* 85: 78–96.
52. Brahim, M., N. Boussetta, N. Grimi, E. Vorobiev, I. Zieger-Devin, and N. Brosse. 2017. Pretreatment optimization from rapeseed straw and lignin characterization. *Industrial Crops and Products* 95: 643–650.
53. Zikeli, F., T. Ters, K. Fackler, E. Srebotnik, and J. Li. 2016. Wheat straw lignin fractionation and characterization as lignin-carbohydrate complexes. *Industrial Crops and Products* 85: 309–317.
54. Gellerstedt, G. 2015. Softwood kraft lignin: Raw material for the future. *Industrial Crops and Products* 77: 845–854.
55. Wu, Y., S. Zhang, X. Guo and H. Huang. 2008. Adsorption of chromium (III) on lignin. *Bioresource Technology* 99: 7709–7715.
56. Guo, X., S. Zhang, and X. Shan. 2008. Adsorption of metal ions on lignin. *Journal of Hazardous Materials* 151: 134–142.

57. Klapiszewski, L., P. Bartczak, M. Wysokowski, M. Jankowska, K. Kabat, and T. Jesionowski. 2015. Silica conjugated with kraft lignin and its use as a novel 'green' sorbent for hazardous metal ions removal. *Chemical Engineering Journal* 260: 684–693.
58. Zhou, C., Q. Gao, S. Wang, Y. Gong, K. Xia, B. Han, M. Li, and Y. Ling. 2016. Remarkable performance of magnetized chitosan-decorated lignocellulose fiber towards biosorptive removal of acidic azo colorant from aqueous environment. *Reactive and Functional Polymers* 100: 97–106.
59. Mussatto, S.I., M. Fernandes, G.J.M. Rocha, J.J.M. Orfao, J.A. Teixeira, and I.C. Roberto. 2010. Production, characterization and application of activated carbon from brewer's spent grain lignin. *Bioresource Technology* 101: 2450–2457.
60. Gwenzi, W., N. Chaukura, F.N.D. Mukome, S. Machado, and B. Nyamasoka. 2015. Biochar production and applications in sub-Saharan Africa: Opportunities, constraints, risks and uncertainties. *Journal of Environmental Management* 150: 250–261.
61. Ge, Y., L. Qin, and Z. Li. 2016. Lignin microspheres: An effective and recyclable natural polymer-based adsorbent for lead ion removal. *Materials and Design* 95: 141–147.
62. Kai, D., M.J. Tan, P.L. Chee, Y.K. Chua, Y.L. Yap, and X.J. Loh. 2016. Towards lignin-based functional materials in a sustainable world. *Green Chemistry* 18: 1175–1200.
63. Nevárez, L.A.M., L.B. Casarrubias, A. Celzard, V. Fierro, V.T. Muñoz, A.C. Davila, J.R.T. Lubian, and G.G. Sánchez. 2011. Biopolymer-based nanocomposites: effect of lignin acetylation in cellulose triacetate films. *Science and Technology of Advanced Materials* 12: 1–16.
64. Shitrit, Y., and H. Bianco-Peled. 2017. Acrylated chitosan for mucoadhesive drug delivery systems. *International Journal of Pharmaceutics* 517: 247–255.
65. Moreno-Vásquez, M.J., E.L. Valenzuela-Buitimea, M. Plascencia-Jatomea, J.C. Encinas-Encinas, F. Rodríguez-Félix, S. Sánchez-Valdes, E.C. Rosas-Burgos, V.M. Ocano-Higuera, and A.Z. Graciano-Verdugo. 2017. Functionalization of chitosan by a free radical reaction: Characterization, antioxidant and antibacterial potential. *Carbohydrate Polymers* 155: 117–127.
66. Gullón, B., M.I. Montenegro, A.I. Ruiz-Matute, A. Cardelle-Cobasa, N. Corzo, and M.E. Pintado. 2016. Synthesis, optimization and structural characterization of achitosan–glucose derivative obtained by the Maillard reaction. *Carbohydrate Polymers* 137: 382–389.
67. Thulluri, C., S.R. Pinnamaneni, P.R. Shetty, and U. Addepally. 2016. Synthesis of Lignin-based nanomaterials/nanocomposites: Recent trends and future perspectives. *Industrial Biotechnology* 12: 153–160.
68. Volkova, N., V. Ibrahim, R. Hatti-Kaul, and L. Wadso. 2012. Water sorption isotherms of Kraft lignin and its composites. *Carbohydrate Polymers* 87: 1817–1821.
69. Naseem, A., S. Tabasum, K.M. Zia, M. Zuber, M. Ali, and A. Noreen. 2016. Lignin-derivatives based polymers, blends and composites: A review. *International Journal of Biological Macromolecules* 93: 296–313.
70. Ma, X., X. Zheng, H. Yang, H. Wu, S. Cao, L. Chen, and L. Huang. 2016. A perspective on lignin effects on hemicelluloses dissolution for bamboo pretreatment. *Industrial Crops and Products* 94: 117–121.
71. Kim, S., M.M. Fernandes, T. Matama, A. Loureiro, A.C. Gomes, and A. Cavaco-Paulo. 2013. Chitosan-lignosulfonates sono-chemically prepared nanoparticles: Characterisation and potential applications. *Colloids and Surfaces B: Biointerfaces* 103: 1–8.
72. Vetrivel, V., K. Rajendran, and V. Kalaiselvi. 2015. Synthesis and characterization of Pure Titanium dioxide nanoparticles by Sol- gel method. *International Journal of Chemical Technology Research* 7: 1090–1097.
73. Feldman, D. 2016. Lignin nanocomposites. *Journal of Macromolecular Science, Part A Pure and Applied Chemistry* 53: 382–387.
74. Bee, A., L. Obeid, R. Mbolantenaina, M. Welschbillig, and D. Talbot. 2017. Magnetic chitosan/clay beads: A magsorbent for the removal of cationic dye from water. *Journal of Magnetism and Magnetic Materials* 421: 59–64.

75. Ngah, W.S.W., S. Fatinathan. and N.A. Yosop. 2011. Isotherm and kinetic studies on the adsorption of humic acid onto chitosan-H$_2$SO$_4$ beads. *Desalination* 272: 293–300.
76. Zeng, D., J. Wu, and J.F. Kennedy. 2008. Application of a chitosan flocculant to water treatment. *Carbohydrate Polymers* 71: 135–139.
77. Zhou, R., R. Zhou, X. Zhang, S. Tu, Y. Yin, S. Yang, and L. Ye. 2016. An efficient bio-adsorbent for the removal of dye: Adsorption studies and cold atmospheric plasma regeneration. *Journal of the Taiwan Institute of Chemical Engineers* 68: 372–378.
78. Richter, A.P., B. Bharti, H.B. Armstrong, J.S. Brown, D. Plemmons, V.N. Paunov, S.D. Stoyanov, and O.D. Velev. 2016. Synthesis and characterization of biodegradable lignin nanoparticles with tunable surface properties. *Langmuir* 32: 6468–6477.
79. Nitayaphat, W., and T. Jintakosol. 2015. Removal of silver(I) from aqueous solutions by chitosan/bamboo charcoal composite beads. *Journal of Cleaner Production* 87: 850–855.
80. Wang, X., and C. Wang. 2016. Chitosan-poly(vinyl alcohol)/attapulgite nanocomposites for copper(II) ions removal: pH dependence and adsorption mechanisms. Colloids and Surfaces A: Physicochem. *Eng. Aspects* 500: 186–194.
81. Mishra, A.K., and A.K. Sharma. 2011. Synthesis of—cyclodextrin/chitosan composites for the efficient removal of Cd(II) from aqueous solution. *International Journal of Biological Macromolecules* 49: 504–512.
82. Okoronkwo, A.E., and S.J. Olusegun. 2013. Biosorption of nickel using unmodified and modified lignin extracted from agricultural waste. *Desalination and Water Treatment* 51: 1989–1997.
83. Klapiszewski, L., K. Siwinska-Stefanska, and D. Kolodynska. 2017. Preparation and characterization of novel TiO$_2$/lignin and TiO$_2$-SiO$_2$/lignin hybrids and their use as functional biosorbents for Pb(II). *Chemical Engineering Journal* 314: 169–181.
84. Khamparia, S., and D.K. Jaspal. 2017. Adsorption in combination with ozonation for the treatment of textile waste water: a critical review. *Environmental Science Engineering* 11 (1): 8. doi:10.1007/s11783-017-0899-5.
85. Choi, C., J. Nam, and J. Nah. 2016. Application of chitosan and chitosan derivatives as biomaterials. *Journal of Industrial and Engineering Chemistry* 33: 1–10.
86. Nair, V., A. Panigrahy, and R Vinu. 2014. Development of novel chitosan-lignin composites for adsorption of dyes and metal ions from wastewater. *Chemical Engineering Journal* 254: 491–502.
87. Szymczyk, P., U. Filipkowska, T. Jóźwiak, and M. Kuczajowska-Zadrożna. 2015. The use of chitin and chitosan for the removal of Reactive Black 5 dye. *Progress on Chemistry and Application of Chitin and its Derivatives, Volume XX*. doi:10 15259/PCACD.20.26.

Cationic Nanosorbents Biopolymers: Versatile Materials for Environmental Cleanup

Sandeep K. Shukla, Rashmi Choubey and A.K. Bajpai

Abstract The present chapter aims to focus on biopolymers-based cationic nanoadsorbents which find numerous applications in removal of toxic metal ions from aqueous solutions. Besides giving an introductory account of the cationic nanosorbents of biopolymers, the chapter discusses various naturally occurring cationic polyelectrolytes like chitosan, cationic cellulose, cation cyclodextrins, and cationic dextran. With the mention of a brief idea of preparation of nanoparticles of these biopolymers, their applications in water remediation have been discussed. The chapter ends with remarks on the current limitations of this field and future prospects and scope for the chemists, environmentalists, and scientists.

Keywords Cationic biopolymers · Nanomaterials · Toxic metal ions · Water remediation

1 Introduction

Biopolymers are naturally occurring polymers comprising of monomer units linked together via covalent bonds to yield a giant molecule. The prefix bio carries the meaning that these polymers are prone to biodegradation and they are normally synthesized by living organisms. These biopolymers are abundant in nature as they are produced or derived from a large number of entities like microorganisms, plants, or trees. Usual chemical synthetic routes can also transform various biological sources like vegetable oils, sugars, fats, resins, proteins, amino acids to biopolymers [1]. In contrast to synthetic polymers which may have simple or random structures,

---

S.K. Shukla
Wainganga Division, Central Wataer Commission,
Nagpur 440006, Maharashtra, India

R. Choubey · A.K. Bajpai (✉)
Bose Memorial Research Laboratory, Department of Chemistry,
Government Model Science College, Jabalpur 482001, Madhya Pradesh, India
e-mail: akbmrl@yahoo.co.in

© Springer International Publishing AG 2018
S. Bhardwaj Mishra and A.K. Mishra (eds.), *Bio- and Nanosorbents from Natural Resources*, Springer Series on Polymer and Composite Materials,
https://doi.org/10.1007/978-3-319-68708-7_4

the biopolymers have specific, well-defined three-dimensional structures with definite properties and these make them active in vivo and provide key functions. For instance, the oxygen will not be carried by hemoglobin if the protein is not folded into a quaternary structure. The biopolymers differ from fossil–fuel polymers in respect of their sustainability when coupled with the property of biodegradability.

It is the structure of the monomer that determines the end properties of the polymer such as ability to withstand temperature, stretchability, barrier to gaseous, vapors, and liquid molecules, stability to chemicals, biocompatibility, biodegradability. It is also reported [2] that these biopolymers are transformable to other derivative biopolymers under specific environmental conditions and microorganisms actions.

2 Classification of Biopolymers

The polyhydroxyalkanoates (PHAs) are regarded as the first biomaterials, discovered in 1925 by the French microbiologist, M. Lemoigne, who observed their accumulation in the form of intracellular substance in a bacterial strain of *Bacillus megaterium* [3]. Based on the nature of synthesis, the biopolymers have been classified into the following four categories.

I. Biopolymers which have been derived from biomass such as starch, lingo-cellulosic materials, protein, and lipids;
II. Those produced from microbial such as the PHAs;
III. Biopolymers which are obtained from agro-resources after chemical treatments such as the polylactic acids or PLAs;
IV. Those biopolymers which are conventionally obtained though chemical synthesis such as aliphatic and aromatic hydrocarbon.

A significant category of biopolymers is cationic biomacromolecules which are obtained either from introduction of some cationic group into the backbone or as pendant group of the biopolymer. The cationic biopolymers exhibit peculiar physicochemical properties that enable them to be transformed into more useful materials suitable for biological applications. The cationic biopolymers have attracted the researchers' interests to explore their optimum utilization in various industrial, technological, and biomedical applications including clinical trials [4].

The biopolymers have, however, been classified on the basis of the method of production and their source also, as shown below:

A: Biopolymers which have been directly extracted or removed from vegetal or animal biomass such as polysaccharides and proteins;
B: Those biopolymers which are produced by chemical synthesis involving renewable bio-based\monomers such as polylactic acid (PLA);
C: Microorganisms originated biopolymers such as polyhydroxyalkanoates, cellulose, xanthan, pullulan [5].

It is known that polysaccharides like starch, cellulose are also coined as biopolymers and accompany growth cycles of all living organism. Another class of these biopolymers is proteins which are frequently used to design biodegradable materials. These biopolymers lack in mechanical strength and, therefore, modified through chemical routes to control degradation and improve their mechanical properties [6].

At the present time due to the strict environmental laws and regulations, various methods of removing the heavy metal ions from effluent streams have been proposed. The various conventional methods developed so far are filtration, chemical treatment, UV radiation, adsorption, distillation, precipitation, ion exchange, electrochemical technologies, etc. The various filtration methods use biosand filters, ceramic filters, charcoal bed, and activated carbon bed. These filters are incompetent in removing organic contaminants, not capable to handle high turbidity and bacteria growth on filter media. Filters require regular backwashing which causes high maintenance cost. In chemical precipitation, metals are removed or eliminated using coagulants such as alum, lime, iron salts, and other organic polymers. The method, however, suffers from the drawback: A large amount of sludge is produced that also contains various contaminants [7].

Among various techniques available for removal of toxic metal ions, the adsorption is the most frequently used approach to decontaminate industrial effluents form the toxic metal ions. Moreover, it is also effective in lowering the operational costs and reducing the equipments size besides enhancing the efficiency of metal ions recovery. Since the process of adsorption works at interfaces, it is essential for toxic moieties to seek high surfaces over the adsorbent materials. The requirement of large surface area is best meet out by the nanoparticles, and therefore they are one of the strongest candidates for adsorption. The discipline of nanoscience has flourished lot in several last decades. The nanoparticles offer high energy adsorption sites as well as greater binding energies for toxic metal ions on their surfaces in comparison with traditional adsorbents. The quality of the effluent generated is also better than the rest of the processes for the reason that the adsorbent has a high affinity toward the metal ions. The affinity may be due to electrostatic forces of the solute to the adsorbent surface, van der Waals attraction, or chemisorptions. It also has a benefit of reversibility, where the adsorptive bed can be regenerated when it gets exhausted with metal ions. This process is mostly preferred as it requires low maintenance cost, has high metal removal efficiency, easy to operate, and uses solid adsorbent which resists degradation [8].

3 Nanoadsorbents

Nanoadsorbents find extensive and variety of applications in science and engineering field as they are efficient biocompatible adsorbents having large specific surface area, more active sites, and low intraparticle resistances. Nanoadsorbents have nanosize pores, high selectivity, high surface area, high permeability, good mechanical strength, and good thermal stability [9]. Nanomaterials may further be divided into four types [10].

- The first type of nanomaterials contains materials based on carbon which are available as ellipsoids, hollow spheres, and tubes. The materials having spherical and ellipsoidal shape are commonly coined as fullerenes, whereas the ones with cylindrical shape are normally called nanotubes and they can remove pollutants from industrial waste water due to the fact that they can establish $\pi-\pi$ electrostatic interactions [11].
- Another class of nanomaterials based on metals comprises of quantum dots, nanogold, nanosilver, and metal oxides, such as titanium dioxide.
- A special class of highly branched organic materials is called as dendrimers which are also coined as nanosized polymers. From structural considerations, the dendrimers have perturbing chains that are responsible for specific chemical functions. This property of dendrimers makes them useful in variety of catalytic reactions.
- Composite nanoparticles are another class of nanoparticles produced by combination of polymers and nanosize clays that results in an enhanced mechanical, thermal, barrier, and flame-retardant properties.

The variety of nanoadsorbents may be proposed that are nanotubes, nanomesh, nanofiltration membranes, nanofibrous alumina filters, magnetic nanoparticles, nanoporous ceramics and clays, cyclodextrin nanoporous polymer, polypyrrole–carbon nanotube composites, etc. [12]. Carbon nanotubes have large specific surface area and small hollow and layered structures making them show potential removal capacity for a variety of organic pollutants and metal ions. They can be easily modified by chemical treatment to increase their adsorption capacity. Nanotubes provide faster flow rates in spite of smaller pores because of smooth interior of the nanotubes. It saves energy as it shows fast flow rates that reduce the amount of pressure required to push the water through the tubes. It can be cleaned by ultrasonification and autoclaving at 121 °C for 30 min and reuse with the same filtering efficiency [6].

A major cause of increasing pollution is widening urbanization and steeply rising industrial activities that are responsible for heavy discharge of toxic contaminants into water bodies that has greatly engendered human life [13]. The uncontrolled discharge of pollutants and contaminants due to their severe toxicity and more carcinogenic nature as comparable to other water contaminants, metal ions can cause brutal health troubles for aquatic fauna as well as flora but it also troubles human health through the prevailing ecological food chain [14]. Thus, the instantaneous removal of these toxic pollutants from wastewater is a noticeable issue in the aerobic and aquatic world. The scarcity of pure drinking water has to be looked upon with two angles: first the quantum of available water and second its quality. The ever increased industrial activities like batteries manufacturing, mining industries, metal plating, agrochemicals like pesticides, herbicides, leather industries have further intensified the problems [15]. It is also known that there are found phosphorous and nitrogen compounds and other toxic heavy metals in the wastewater and sludge due to increasing domestic uses of plumbing, healthcare products, surfactants, and other large-scale commercial activities [16].

Normally in wastewater, there are found a variety of entities like soluble compounds and colloidal suspensions having dimensions from 0.001 to 10 µm. The presence of these impurities in water results in an increased, color, turbidity, chemical oxygen demand of etc. [17].

There has been a noticeable rise in industrial activities that has greatly contributed to enhanced pollution. The most common contaminants are organic compounds, aromatics, halogenated complex compounds, heavy metals, etc., and they have severely affected both the human life and the environments also [18]. The heavy metal contaminations are more dangerous as these metal ions are totally nonbiodegradable and persist in aquatic and environmental systems for longer times even up to several months and years also. Their persistence causes severe physiological effects like prolonged illness, development of carcinogenic tumors, chronic indigestion problems. [19]. Among various metal ions, the ones that are most frequently and severely responsible for health-related issues are zinc (Zn), copper (Cu), nickel (Ni), mercury (Hg), cadmium (Cd), lead (Pb), and chromium (Cr) [20].

Nanosorbents are nanoscale particles or nanocomposites that have high affinity to absorb substances from organic/inorganic solution due to small size and high surface area. Nanoadsorbants can be designed to have multiple reactive nanostructure components to target specific contaminants and propose the opportunity of even greater sorption capacities [21]. Nanosorbents have application in air or water purification and also in remediation of groundwater or wastewater treatment process. They have a large surface area available for reacting with the pollutants, and their small size provides more mobility, so they can be transported effectively by the groundwater. Apart from this, nanoparticles maintain their properties for a long time and never affected by soil acidity, temperature, or nutrient levels and they can be efficiently applied for ex situ or in situ treatment for wastewater at a field scale. Hence, remediation technology is significantly improved in terms of efficiency, selectivity, and specificity when compared to conventional technologies. Selection of the best method and materials for water purification must follow four conditions: (1) How far the treatment is flexible and efficient, (2) reusability after adsorption, (3) environmental friendliness, and (4) economic viability.

3.1 Chitosan

This biopolymer is basically derived from chitin which constitutes the second most abundant biopolymer found in environment and present in crystalline microfibrils forming structural components. The chitosan offers great potential to adsorb large number of heavy and toxic metal ions and even radio nucleotides [22]. Chitin was first isolated by Braconnot in the year 1811 [23] as fungi, and later in 1823, it was discovered by Odier in insects and named as chitin. At present, it is produced from shrimp and crab shells which are obtained from the seafood industry. These shells contain 30–50% calcium carbonate, 30–40% protein, and the rest 20–30% is chitin (dry mass) [24].

The properties like positive charge, compatible and biodegradable nature, non-toxicity, antibacterial character, favorable physical and chemical behavior make the chitosan a versatile material that finds exhaustive applications in medicine, pharmacy, food industries, cosmetics, agriculture, water remediation, biomedical applications like drug delivery, wound dressings, artificial implants [25].

Structure of chitin and chitosan: As opposed to chitosan, the chitin has totally different properties like, it is a hydrophobic, stiff, and inelastic material found as white and un-reactive solid, and does not dissolve in water and common organic solvents. However, it is soluble in hexafluoroisopropanol, hexafluoroacetone, and chloroalcohols in synthesis with aqueous solutions [26]. It is a homopolymer made of linear structural units of β-(1, 4)-lined 2-acetamino-2-deoxy-β-D-glucocopyranose with 2-amino-2-deoxy-β-D-glucopyranose (Fig. 1). The presence of large number of amino and hydroxyl groups ensures its metal ions uptake potential either through chelation or ion exchange pathway when this biopolymer comes in contact with metal ions solution [27]. The properties of chitosan such as its molecular mass, degree of purity, crystallinity, and surface morphology depend on the sources. The chitin is normally found in three crystalline polymorphic forms (α-, β-, or Υ-chitin) defined on the basis of crystalline chain packing regions [28]. The α- form consists of antiparallel chains, β- form has two parallel chains (stack structure), and Υ- has one antiparallel chain in the structure. A similar structural feature of α- and β-chitins is that in both of the forms, C=O···H–N intermolecular hydrogen bonds are present, whereas the –CH_3OH groups are present only in α-chitin but not in β-chitin. Because of these hydrophilic groups, the β-chitin has affinity to water molecules and it swells thus producing hydrates. This feature is missing in α-chitin. The β-chitin is more prevalent in squid and marine diatoms which occur rarely, whereas the α-chitin is mostly found in crustaceans, insects, and fungi. Chitosan can be conveniently obtained from most available α-chitin [29].

The simple way to prepare chitosan is the N-deacetylation of chitin, and the prepared chitosan easily dissolves in acetic acid to give a clear solution that upon casting results in a free standing film [30]. Chitosan generally prepared from most available α-chitin [31]. Chitin is known to be a semicrystalline biopolymer that contains biopolymer chains bonded to one another via intra- and intermolecular hydrogen bonds, thus imparting insolubility character to chitin in organic solvents and diluted acids. When the degree of deacetylation exceeds over 60%, the chitin easily dissolves in dilute acidic solutions and this kind of chitin is termed as chitosan. In essence, chitosan is a biopolymer derived from deacetylated chitin, and due to its outstanding physical, chemical, and biological properties, this biopolymer finds comprehensive applications [25].

People pay more attentions to chitosan because it has very strong adsorption capacity of heavy metals and low cost. Some Asian countries such as Japan, Thailand, China's fishery wastes such as shrimp shell, crab shells were used to produce chitosan. Because the fish waste source is rich, chitosan is also cheap. The yield of chitin is one of the world's second-largest biological polymer, second only to cellulose. Chitosan is more important than the chitin, and the molecular structure is similar to cellulose. At present, the chitosan attracted much attention due to its adsorption ability is strong. It mainly exists in the exoskeletons of crustaceans. In recent years, many researchers study on a large amount of chitosan [32]. At commercial level, the chitosan is normally prepared by using NaOH as deacetylating agent of chitin; however, enzymatic conversion routes are also reported [33].

There are several processes involved in the conversion of carb of shrimp shells to chitin such as (a) chemical or enzymatic desprotenization, (b) demineralization by acidic treatment to remove calcium carbonate and other minerals, and finally (c) discoloration to remove residual pigments.

Chitosan biosorption: The degree of crystallinity is the key factor that decides the metal ion sorption capacity of chitosan, its affinity for aqueous solutions, and the extent of deacetylation. In comparison with chitin, the chelating ability of chitosan is five- to sixfolds greater than that of the chitin which is basically due to the presence of free amino groups of chitosan. These characteristics of chitosan make it a suitable candidate for cleaning up of environment, water remediation, and separation and recovery of precious metal ions. However, the processing of chitin to chitosan increases the material cost and restricts its acceptance for industrial utility. Furthermore, the availability of chitosan with different characteristics and sources also discourages its industrial acceptance [34]. The exhaustive applications of chitosan in removal of metal ions from aqueous solutions are owing to its hydrophilic groups also that facilitate diffusion of chitosan chains in aqueous solutions of metal ions and thus cause rapid adsorption of metal ions from solutions or effluents. Furthermore, the hydroxyl and amino groups of chitosan can also react with solutes present in solution. However, from the adsorption point of view, the amino groups are more important than the hydroxyl groups and it is the amino groups only that decide the quality of this biopolymer [35]. In spite of good adsorptive potential of chitosan, there are some limitations like dissolution of chitosan in acidic solution of metal ions that limits its utility especially in acidic industrial effluents or water

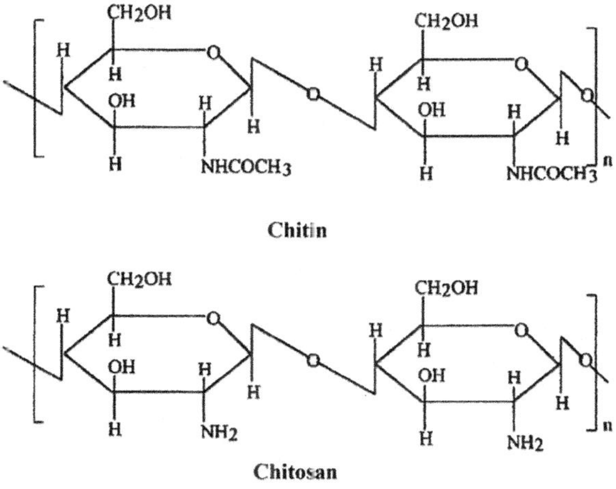

Fig. 1 Chemical structures of chitin and chitosan

solutions of highly acidic nature [36]. A remedy of this problem of dissolution of chitosan is its enhanced cross-linking but it often results in suppressed adsorption of metal ions.

The chitosan contains high nitrogen on its molecules, and they really work as active sites to cause various chemical reactions in aqueous solutions. These amino groups of chitosan are not strong enough to deprotonate water according to the following reaction scheme [37].

$$Chitosan - NH_2 + H_2O = Chitosan - NH^{3+} + OH^-.$$

The high affinity of chitosan for metal ions has motivated researchers to develop new and newer adsorbents based on chitosan and its derivatives for effective adsorption and removal of metal ions from aqueous solutions [38]. The removal of heavy metal ions has been reported using chitosan-based membranes [39]. In a study by Liu and Bai [40], fabrication of porous and hollow fiber membranes of chitosan and cellulose acetate (CA) was reported for removal of copper ions from water. It was also noticed that when the chitosan content was high, the membranes exhibited greater affinity for the copper ions. The researchers suggested adsorption-assisted mechanism for the removal of metal ions and water treatment. In another study by Salehi et al. [41], the fabrication of chitosan/PVA thin adsorptive membranes embedded with amino functionalized multiwalled carbon nanotubes was proposed for adsorption of copper ions. It was also noticed that greater adsorption of copper was obtained when the nanotubes were enriched in carbon and moreover when the temperature was high.

The researchers also introduced poly (ethylene) glycol (PEG) into the membranes to make it macroporous for enhanced adsorption of metal ions. It was also found that 5 wt% of PEG was enough to impart mechanical stability to membrane and improve its adsorptive capacity [38]. It was also noticed that if multiwalled carbon tubes are impregnated into the membrane, a greater sorption capacity and mechanical stability may be achieved even in static and flow conditions. The inclusion of PEG was found to cause generation of interconnected pores and nanochannels in the membrane. The membrane offered adsorption sites due to both the amino groups and MWCNTs. The formation of coordination bonds between the functional groups of chitosan and added filler has been schematically presented in Fig. 2 [42].

Upon contact with water, the chitosan results in an increase of pH due to its pKa of 6.3 [43]. The adsorptions on chitosan are pH-dependent process as a consequence of acid–base reaction. The deprotonated amino groups of chitosan function as active sites for adsorption of metal ions via chelation mechanism. Moreover, the electrostatic properties of chitosan may also be a mechanistic route for adsorption through ion exchange mechanism [44].

Chitosan and alginate have been separately investigated as adsorbents for metal ions like Cu(II), Cd (II), Pb(II), Ni(II), Hg(II), Cr(VI), U(VI), Mo(V), V(V), Pd(II), Pt (IV), Au(III), As(V), Se(V) showing high adsorption capacities [45]. The maximum

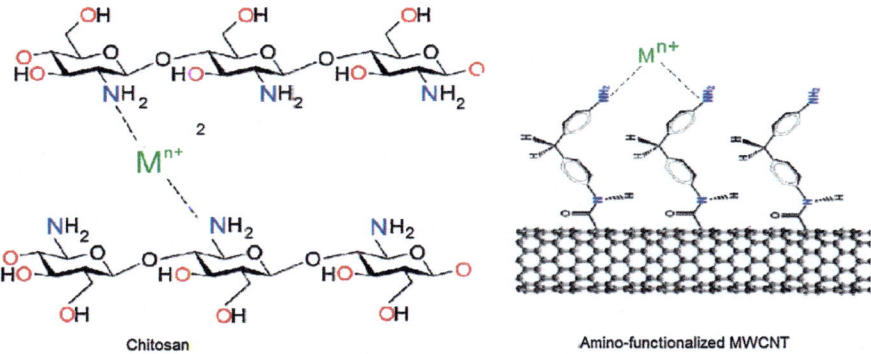

Fig. 2 Adsorption of metal ions on chitosan through coordination with amine groups

adsorption capacity depends on factors such as the forms of chitosan and alginate (beads, powder), experimental conditions like pH, temperature, solution composition, nature of modification of chitosan, i.e., cross-linking, grafting, etc. [46].

3.2 Cationic Cellulose

Cellulose, the most abundant biological material on the Earth, deserves well to cater the needs of products which are biocompatible and environmentally friendly. From compositional point of view, it is made up of glucose units joined through beta glycosidic bonds. This gives rise to densely packed straight shape macromolecules which impart outstanding mechanical strength to the cellulosic materials like the wood [47]. The cellulose is also present as a constitutive component in plant cell walls. The cotton is regarded as the purest form of cellulose and has great applications in textile industries [48].

If one looks at the chemical structure of cellulose, it is made up of long chains comprising of six-membered ring glucose molecules. Two anhydroglucose rings (C6H10O5) n, where $n = 10{,}000$–$15{,}000$ forms the repeat unit of cellulose. Cellulose is composed of fibers which are not indigested by humans but some animals digest it [49]. A lot of biopolymer composites are made up of cellulose only which are considered as an idea material. For example, one of the most frequently used derivatives of cellulose is carboxymethyl cellulose (CMC) which has been employed in the preparation of silver nanoparticles. It is found that the properties of silver nanoparticles such as size and stability greatly depend on the degree of polymerization of CMC which also determines its reduction capacity [50].

As shown in Fig. 3, the basic chemical structure of cellulose is a linear homopolysaccharide composed of β-D-glucopyranose units linked together by β-1-4-linkages [51]. The chemical structure clearly reveals that there are present three OH groups per molecule. Thus, it is now very clear that the presence of three

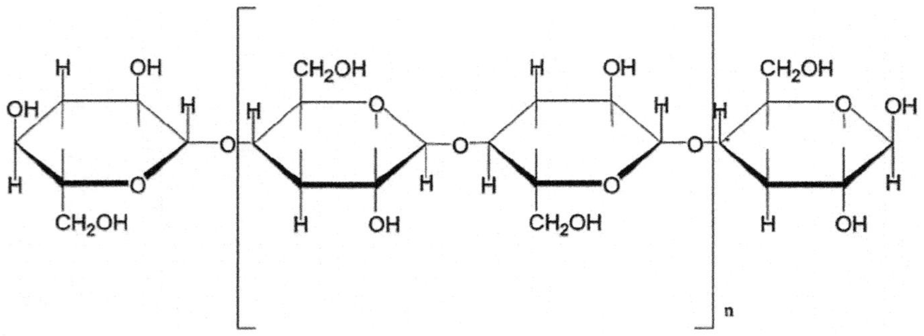

Fig. 3 Structure of cellulose showing the cellobiose repeat unit

OH groups facilitates formation of hydrogen bonds which, in turn, plays a key role in arrangement of crystalline packing and consequently dictates the physical properties of cellulose including mechanical strength [52]. The physical forces like hydrogen bonding and van der Waals force glucan chains to pack side by side and form cellulose microfibrils which are then stacked to form crystalline cellulose [53]. It is clear that the –OH groups which are quite polar and lie along the neighboring chains tend to result in bundling of the chains. The regular stacking and bundling of chains form hard and stable crystalline regions, which provide strength and stability to the cellulosic materials. The length of the chain greatly varies depending on the nature of compound. For instance, there are only few hundred sugar units in wood pulp while more than 6000 in cotton [54].

Recently, the functionalization of cellulose derivatives has gained much research interests that aim at expanding the application domains of cellulose. These research trends permit to keep together various functional groups with desired composition to achieve favorable properties. One of the important forms of cellulose is cationic cellulose which is water soluble and has a prime place in smart materials due to its pH and ionic strength-dependent behavior. It is reported that the cationic cellulose has potential to find application in drug delivery technology [55]. Besides biomedical applications, the cationic derivatives also find applications as flocculants in wastewater treatment, mineral processing and oil recovery, and many more domestic and healthcare products [56].

Functionalization of cellulose is another active area of synthetic chemistry that often results in cellulose derivatives with outstanding applications [57]. Amination of hydroxyethyl cellulose (HEC) produces water soluble cationic cellulose derivative which contains 4% coupled nitrogen. For this purpose, two etherification agents, diethylepoxypropylamine (DEEPA) and (2-chlorethyl) diethylamine hydrochloride (DEAE) [58], were used. The product formed was aminocellulose which was characterized by the techniques like FTIR and Raman spectroscopy. The reactions of HEC with DEEPA and DEAE result in a cationic polyelectrolyte (pKa approx. 8.2) and polyfunctional with pKa 6.0–6.2 and 8.2–8.6, respectively. The prepared derivatives dissolve in polar aprotic solvents and water. Furthermore, the solutions of the prepared derivatives have greater stability at high temperature and high salt concentrations. In another work, the hydroxypropyl cellulose (HPC) was also functionalized to prepare its derivatives which were characterized by different analytical techniques including the ^{1}H NMR spectroscopy and elemental analysis [59].

One of the recent research activities in cellulose chemistry includes preparation of cellulose nanofibres (CNF) which are produced by grinding of cellulose or its high-pressure fluidization so as to remove the present lignin content. The cellulose nanofibres are 5–20 nm thin and several μm long fibrils having high aspect ratio. When the concentration of cellulose is low, then a transparent gel is formed which can be used to produce films which are biodegradable and environmental safe and may be employed for different applications. There are available several sources like coir, banana, sugar from which the cellulosic nanofibres may be extracted. To these extracts, some suitable plasticizers may be added that substantially improves the physical, chemical, and mechanical properties of the end products. The purpose of adding plasticizer to the cellulose nanofibres is to provide grease proofness and high barrier against passage of oxygen under dry conditions. It is known that cellulose synthesizing organisms such as bacteria, algae, tunicates, and higher plants have cellulose synthase proteins, which catalyze the polymerization of glucan chains [59].

3.3 Cationic Cyclodextrin

These are naturally occurring cyclic oligosaccharides discovered more than 100 years ago [60]. However, it has been made possible recently only to achieve highly purified cyclodextrins which are currently being used in pharmaceuticals. At present, there are more than 30 pharmaceutical products which are available globally at commercial level. One of the major applications of these cyclodextrins in pharma industry is to enhance solubility of drugs which are normally not soluble in water [61].

3.3.1 Structure and Properties

The molecules of cyclodextrins consist of cyclic oligosaccharide having outer surface of hydrophilic nature. They comprise of (α-1,4-)-linked α-D-glucopyranose units with a lipophilic cavity at center (Fig. 4). The chair conformation of the glucopyranose units makes their shape like cones with secondary hydroxy groups extending from the wider edge and the primary groups from the narrow edge. These structural features make the whole cyclodextrin molecules with hydrophilic outer surfaces, while the lipophilic cavity at the center mimics like an aqueous ethanolic solution [62]. The frequently found natural cyclodextrins are made up of six (α-cyclodextrin), seven (β-cyclodextrin), and eight (γ-cyclodextrin) glucopyranose units. The β-cyclodextrins have limited solubility in water which reveals that when they interact with lipophiles the formation of complexes occurs resulting in precipitation of solid cyclodextrin complexes. This could possibly be due to strong binding of cyclodextrin molecules in the crystal state those result from high crystal lattice energy [63]. It has been found that when the hydroxyl groups are randomly substituted by hydrophobic moieties, their solubility is significantly improved. The enhanced solubility may be attributed to the fact that substitution of hydrophobic groups tends to change the cyclodextrin from crystalline to amorphous nature. Table 1 presents various derivatives of β-cyclodextrins. Basically, the molecules of cyclodextrin are quite large having molecular weight ranging from 1000 to more than 2000 Da. Moreover, due to the presence of large number of hydrogen donors and acceptors, their absorption through biological membranes is quite poor [64].

Cyclodextrins have cage-type shape and resemble supramolecular structure. They possess chelating property and have strong adsorbing nature [65]. The chelating property of β-cyclodextrin can be substantially improved by several chemical modification processes like esterification, etherification, oxidation reactions, and cross-linking of hydroxyls outside the interior cavity [66]. They also have a strong property to adsorb metal ions.

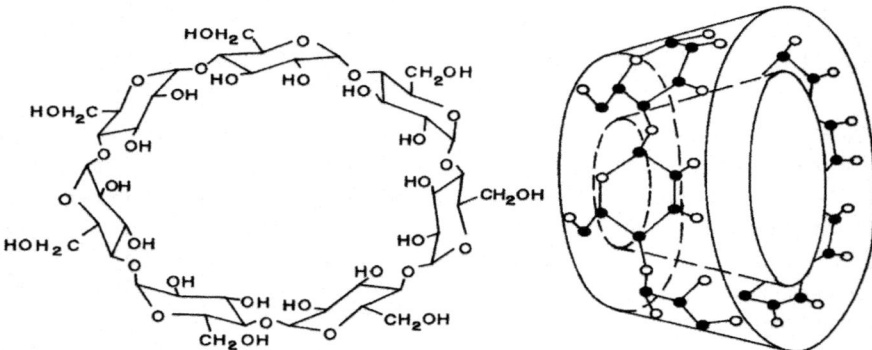

Fig. 4 Chemical structure showing the conical shape of the β-cyclodextrin molecule [reproduced with permission from J. Pharm. Sci. (1996) 85: 1017–1025]

Table 1 Various derivatives of cyclodextrins [67]

Cyclodextrin	R = H or
β-Cyclodextrin	–H
2-Hydroxypropyl-β-cyclodextrin	–CH$_2$CHOHCH$_3$
Sulfobutylether β-cyclodextrin sodium salt	–(CH$_2$)$_4$SO$_3$–Na$^+$
Randomly methylated β-cyclodextrin	–CH$_3$
Branched β-cyclodextrin	Glucosyl or maltosyl group

The major structural features of CDs include their monodisperse saccharide structure, abilities to get desirably modified through chemical reactions, and almost nil toxicity. It is interesting to note that oligosaccharides have low immunogenicity and possess multiple sites through which cationic or cell targeting moieties may easily be introduced. Furthermore, the cationic CDs exhibit a great tendency toward binding of nucleotides and consequently augmented delivery by viral vectors. The incorporation of polycationic polymers and dendritic vectors into CDs has already been achieved which play a key role in therapeutic applications [68].

The typical structure of β-cyclodextrin (β-CD) [69] makes it materials of first choice in many biomedical and environmental applications. Moreover, renewable, biodegradable characters and large number of active hydroxyl groups selectively make cyclodextrin a promising adsorbent for Cu (II) ion removal from wastewaters [70]. Zhao et al. chemically modified cyclodextrin by esterification and oxidation reactions and cross-linked OH groups lying at the exterior of the central cavity to increase their adsorption capacity toward toxic heavy metal ions [71].

Cyclodextrin adsorbent CDA is a co-polymer resin formed from acrylic acid (AA) and acrylamide (AM) by inverse suspension and redox titrations to remove Cu (II) ion from wastewater capacity when the pH of the solution and the ionic strength increased was 107.37 mg/g at 80 mg/l concentration of Cu (II) ion. Experimental data fitted best with the Freundlich equation model. The adsorption kinetics data revealed that Cu (II) ions adsorb onto the CDAA following theoretical models of quasi-second order and Elovich equations [72]. Allabashi [73] reported the synthesis and characterization of triethoxysilylated derivatives of poly(propylene imine) dendrimer, polyethylene imine and polyglycerol hyper-branched polymers and beta-cyclodextrin and impregnated ceramics into the modified materials to design membrane filters which were demonstrated to show enormous potential for removal of variety of organic pollutants from water. The fabricated filters were found to show extremely high removal of aromatic hydrocarbons, trihalogen methane, pesticides, and methyl tert-butyl ether.

The cyclic oligosaccharides, namely a-, b-, g-cyclodextrins (CD), are significant materials due to the ability of their cavity to encapsulate hydrophobic drugs. The encapsulation capacity greatly depends on the number of glucose units [74]. There are so many biomedical applications in which cyclodextrin has been grafted on to polysaccharides to produce biocompatible materials. It was also reported that a- and b-cyclodextrin chitosan may be achieved with higher substitution and the prepared derivative so cyclodextrin has ability to distinguish and identify specific guest molecules depending on their structures and shapes [75]. The synthesized polymers were suggested for applications in adsorption and drug delivery systems.

Ozmen and Yilmaz [76] reported the synthesis of two starch and beta-cyclodextrin-based polymers and evaluated their potential as adsorbents for removal of Congo red from aqueous solutions using batch and column procedures. The effect of various experimental conditions was also investigated on the extent of dye removal. It was noticed that the pH of the solution has strong impact on the dye removal capacity of the adsorbent, and at pH 7.0 the removal was found to be the optimum. The adsorption data were fitted best to the Freundlich equation.

The cyclodextrin good has ability to form inclusion complexes with organic and inorganic compounds in the central cavity [77]. It was demonstrated by Hu et al. that after grafting of cyclodextrin onto the multiwall carbon nanotubes/iron oxide better results of lead removal were obtained due to strong complexation of lead ions with the OH groups of cyclodextrin [78].

3.4 Cationic Dextran

Dextran is known to be a naturally occurring branched polymer offering excellent biodegradability and biocompatibility. This polymer has been approved by the FDA and shows high solubility in water. Dextran has wide spectrum of applications in environmental protection as well as biomedical fields. The cationic derivatives of dextran have also been used in drug delivery of nucleic acids [79].

3.4.1 Structure and Properties

The molecule of dextran is made up of glucose units, and this polymer contains chains of different lengths. As shown in Fig. 5, the molecule of dextran contains large number of α (1→6) glucosidic linkage and a variable amount of α (1→2), α (1→3) and α (1→4) branched linkages [80] as shown in Fig. 5. The source of bacterial strain determines the degree and type of branching. The average molecular weight of this polymer lies in the range 107–108 Da [81] but the molecular weight can be reduced to smaller fractions by performing its acidic hydrolysis.

These properties are due to its water solubility irrespective of the pH of the solution and of its polysaccharide structure consisting of α1,6-linked glucose units as well. Utilizing a debranching enzyme, living organisms are able to break this chemically stable bond. The three accessible hydroxyl moieties, found on every

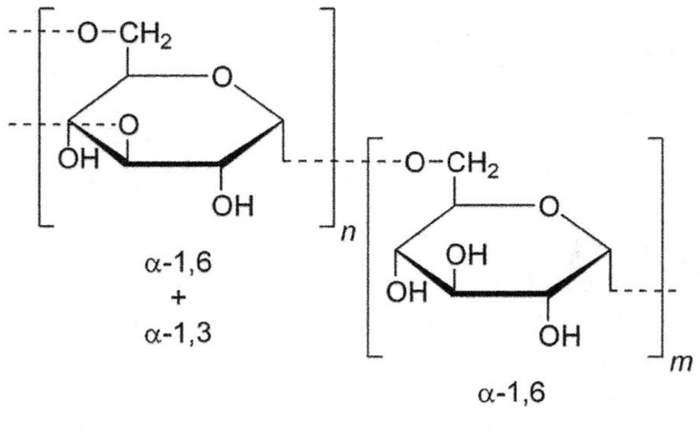

Fig. 5 Chemical structure of dextran

monomer unit, facilitate modifications such as the incorporation of amino groups. (Diethylaminoethyl) dextran [82] and dextran–spermine [33, 84] are well elucidated examples, especially since dextran–spermine shows high transfection efficiency for DNA [85].

The polyglucans are synthesized from sucrose by many species of the genera Leuconostoc, Lactobacillus, and Streptococcus. Hucker and Pederson [86] were the first to disclose that it is the sucrose molecule that produces detxtran by stains of Leuconostoc species. Jeans et al. [87] explained how different strains of bacteria may produce dextran. It was also reported that there are other bacteria also which can produce dextran. There is available soluble and insoluble type of dextran having molecular weights ranging from 1.5×10^4 to 2×10^7. Owing to wide industrial applications, the dextran is produced at commercial scale using the strains of *Leuconostoc mesenteroides* NRRLB–512F [88].

The cationic derivatives of dextran were prepared by substituting hydroxyl groups with glycidyltrimethylammonium chloride (GTMAC) as shown in Fig. 6. The modification of dextran may be done by reacting its hydroxyl groups to other molecules of different properties [89]. By judicious selection of reacting molecules and nature of the grafting species, a variety of amphiphilic dextran derivatives have been synthesized which offer a wide variation in their end properties, and therefore they may find diverse applications depending on their characteristics [90]. Furthermore, there are other variants also like molecular mass, type, and degree of branching that also regulate the properties of derivatives.

Dextran and their derivatives have exhaustive applications in diversified areas like pharmacy, clinical diagnostics, chemical and food technology. A typical biomedical use of dextran is like blood plasma volume expander, drug carrier, and emulsifier and in food industry also. Furthermore, the use of dextran in purification of proteins is also reported that works on the principle of size exclusion chromatography [92]. The derivatives of dextran also have potential applications depending on their properties. A significant use of dextran and its derivatives is in formulating drug delivery systems [93]. The conversion of dextran into nanoparticles has opened up new avenues in designing novel drug delivery systems [94]. Thus, it may be conclusively stated that the dextran and its derivatives are themselves constitute a class of significant materials for vital applications.

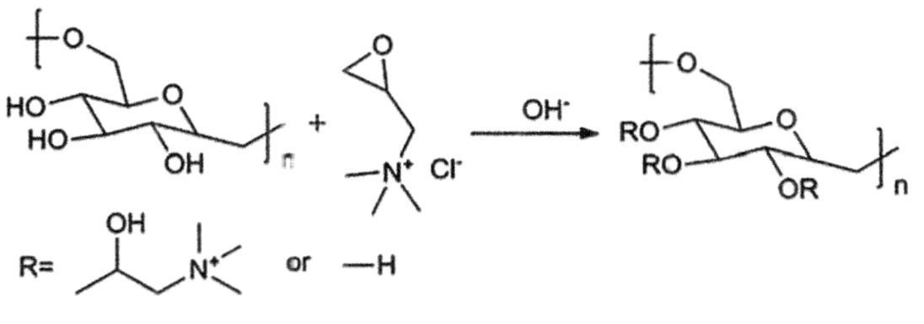

Fig. 6 Reaction scheme showing mechanism of developing a cationic dextran derivative (reproduced with permission [91])

4 Preparation of Cationic Bionanosorbents

Chitosan nanoparticles have been extensively utilized for water remediation purposes. One of the most popular methods include microemulsion cross-linking methods which consist of cross-linking a chitosan emulsion by some suitable cross-linker like glutaraldehyde [95]. A simple and facile technique of preparation of chitosan nanoparticles is by cross-linking this cationic biopolymer with anionically charged multifunctional molecule such as tripolyphosphate which is commonly coined as TPP. Due to nontoxic and multiple charged nature of TPP, it has been widely used in preparation of chitosan nanoparticles of different shapes and sizes. The interaction of chitosan and TPP is termed as ionotropic gelation technique, and it can be precisely controlled by experimental conditions like pH, concentrations of chitosan, and cross-linker, temperature. Nasti et al. [96] investigated the effect of several experimental factors like pH, concentration, ratios of components, and method of mixing, on the preparation of chitosan/TPP nanoparticles. In a typical study by Lin et al. [97], a correlation was established between the number of free amino groups on the chitosan molecule and concentration of the cross-linking agent.

5 Cationic Nanosorbents in Water Remediation

Technologies to seek safe drinking water obtained through effective purifying mechanisms have shown a vast development in the field of nanotechnology. Antibacterial and antifungal properties of various nanomaterials have already been proved. Development of low-cost antibacterial materials like silver nanoparticles which can release constantly to water is an adequate way for providing water which is microbially safe to drink. Several nanocomposites have been developed consisting of functional materials through which the antibacterial agents can release and remove several toxic materials from water such as arsenic, lead, and thus produce purified water that can be affordable without use of electricity [48]. The major difficulty in this technology is the development of stable materials which can constantly release nanoparticles to overcome the scaling results due to the presence of numerous critical species inside water. Performance of comparatively new biopolymer like chitosan is better than conventional polymers. Along with versatile properties, this polymer has wide applications in biomedical fields, water treatment, and dietary supplement industries. The chitosan has been used as a flocculant in the processes of purification of water and degrades frequently in the environment within a weeks or months. Besides chitosan, there are flocculants available that are aggressive and cheaper but their residual impacts remain in the environment. Removal of metals from water by chitosan results due to the formation of the chelates. Chelation is a technique of removal of metal from a solution in the form of cage-like structure produced by the binding of metal with multiple binding sites along the polymer chain. The chelation and biodegradable properties of chitosan make it significant over the conventional method for the treatment of industrial storm water and wastewater, to decline level of contamination. Porous GO–biopolymer gels can efficiently remove cationic dyes and heavy metal ions from wastewater [98]. Prepared granular composite materials of nanocrystalline metal oxy-hydroxide-chitosan via aqueous route efficiently helped in purification of water [99]. Process of water filtration can be carried out by the membranes of nanofibres which do not have any harmful effects on environment. Combination of biopolymers with several nanomaterials can effectively inhibit the formation of biofilms on the surface of polymer.

There are several techniques available that can scavenge toxic metals from effluents including thermal, biological, and chemical methods such as condensation, chemical precipitation, solvent extraction, electrolysis, separation through ultrafiltration membrane, adsorption, irradiation, and electro dialysis. But process of adsorption and ion exchange that comes under technique of sorption is possible substitutes for wastewater purification. Adsorption is a process where a substance (adsorbate) attaches (adsorbs) to another substance (adsorbent) by physical and chemical interactions. The sorbate must diffuse from water or gas phases into the sorbent surface and frequently into the internal pores of sorbent. Removal of heavy metal ions from water with the help of wide use of chelating resins is possible due to their properties of high adsorption capacities, selectivity, and durability [100]. It combines ion exchange and complex formation in its separation process and shows great advances in extraction when compared to conventional resins. Another prominent example of conventional sorbents for heavy metal removal is activated carbons, clay minerals, chitosan/natural zeolites [101]. However, cost and efficiency limits their application at a large scale.

Long alkyl chain functionalized derivatives, like poly (propylene imine) dendrimers, poly (ethylene imine) hyper-branched polymer, and β-cyclodextrin, which are hydrophobic, have property to remove organic pollutants from water [102]. Impregnation of ceramic porous filters with these compounds leads to the formation of hybrid organic/inorganic filter modules. Testing of these hybrid filter modules has been performed for an adequate purification of water, by continuous filtration experiments, employing a different type of water pollutants. It has been established that representatives of the pollutant group of trihalogen methanes (THMs), monoaromatic hydrocarbons (BTX), and pesticides (simazine) can also be removed (>80%), although the filters are saturated considerably faster in these cases.

The adsorption of lead metal ions from aqueous solutions by chitosan nanoparticles was studied by Hejri [103]. Nanoparticles were synthesized by maleic acid from chitosan cross-linking and then analyzing their particle size by the method of dynamic light scattering (DLS). The average particle size of the nanoparticles which can be identified by DLS study was obtained in the range of 70-350 nm. At room temperature (20 ± 0.5 °C), the experiments were performed to observe the effects of the variables like: pH of initial solution, absorbent dosage, and initial concentration of lead (II). Under these surroundings, the optimal values obtained at 6, 55 mg/l and 4 g/l, respectively, for initial pH solution, concentration of metal ions, and adsorbent dose. At 28.5 mg/g, the maximum adsorption was obtained based on optimal values. These adsorption data completely obeyed Langmuir and Freundlich isotherms. On the basis of Langmuir isotherm, a value of 27.35 mg/g as the maximum adsorption rate was obtained. Comparing the analysis of nanoparticles with chitosan polymer particles represents an increase in the effective surface of nanoparticles which increases the efficiency of absorption process, partly, and decrease the costs and the duration of the process, greatly.

Wastewater management has wide range of application of native dextran. Dextran exhibits several favorable properties like stable alkali and acids at room temperature. It is having degradation property and can bind metal ions at alkaline pH. Its use was economically favorable. In the process of flocculation, it is effectively utilized in wastewater treatment [104].

Al-Aidy El-Saied [105] used carboxymethyl–beta-cyclodextrin, poly(ethylene glycol), beta-cyclodextrin, and their magnetic counterparts for discharge of lead and copper from water. FTIR, TGA, and XPS analysis confirms the grafting of CM-b-CD and PEG-b-CD onto the magnetic nanoadsorbents. The solution pH greatly influenced the adsorption of metal ions. The optimum temperature of the solution for attaining maximum adsorption was 45 °C. Maximum adsorption of metal ion was reached after 45 min. It was found that the CD modified with PEG is more efficient that those modified with carboxymethyl because the metal affinity of PEG chain is greater than the metal affinity of the carboxymethyl group. These magnetic nanoadsorbents were used to effectively remove Cu^{2+} and Pb^{2+} from aqueous solution.

In the flocculation-aggregation treatment of municipal effluents, two anionic nanocelluloses (dicarboxylic acid, DCC, and sulphonated ADAC) were tested as flocculants, while model kaolin clay suspensions were used for the study of property of flocculation of cationic nanocellulose (CDAC) and nanocelluloses obtained from sulphonated wheat straw pulp fines (WADAC) to test the lead adsorption [106]. Along with ferric coagulant, the anionic nanocelluloses (DCC and ADAC) performed more efficiently in sewage water treatment by a combined coagulation-flocculation process. Combined treatments for both anionic nanocelluloses can reduce residual turbidity and COD in a stable suspension with high decrease in total chemical consumption compared to coagulation with ferric sulfite alone. Similarly, powerful coagulation of colloidal kaolin produced by CDACs which also efficiently maintain flocculation at variable pH and temperature ranges. The commercial adsorbents and the nanofibrillated and sulphonated fines cellulosic (WADAC) show comparable capacities.

Dubey et al. [107] synthesized biopolymers (chitosan and alginate)-based nanoadsorbents and employed them for the adsorption of mercury ions in aqueous solution. Results of adsorption isotherm show that the process of adsorption is clearly explained by Langmuir model ($R^2 = 0.96054$) in comparison with Freundlich, Temkin, and Dubinin–Radushkevich models. The adsorption of mercury ions is exothermic and can be revealed by the temperature studies. The analysis explained that the produced biopolymer nanomaterials could be efficient and economically viable adsorbent for Hg (II) ions removal. Moreover, the nanoparticles can be recycled subsequently for the removal of metals.

The removal of heavy metals such as zinc (II), nickel(II), copper(II), cobalt(II), and cadmium (II) ions from an aqueous solutions was determined by using succinic anhydride modified mercerized nanocellulose [108]. The FTIR and SEM studies were used to characterize the modified adsorbents. The pH effect, contact time, regeneration, and the concentration of metals were studied in batch mode. The range of maximum uptake of metal from 0.72 to 1.95 mmol/g and its order are the following: Cd > Cu > Zn > Co > Ni. With the help of Langmuir and Sips models with wet and dry weight of adsorbent, an adsorption isotherm was explained. Together, these models were representative to simulate adsorption isotherms. Modified nanocellulose was regenerated and accomplished using nitric acid and ultrasonic treatment.

The nanostructured aminopropyltriethoxysilane (APS) modified microfibrillated cellulose (MFC) adsorbent was synthesized by Hokkanen et al. [109] for removal of heavy metal ions from aqueous solutions. The synthesis involved no hazardous chemicals or processes. The prepared adsorbent was quite efficient in removing Ni (II), Cu(II), and Cd (II) removal from contaminated water. The adsorption of Ni(II), Cu(II), and Cd (II) ions was shown to be dependent on the solution pH. The kinetic study demonstrated that the kinetic mechanism for the adsorption of metal ions followed a pseudo-second-order model, which provided the best correlation with the experimental data. Also the intraparticle diffusion model was well fitted. In the isotherm studies, the experimental maximum adsorption capacities ranged from 2.72 to 4.20 mmol/g. The Sips isotherm model provided the best fit to the experimental adsorption data for these ions, revealing maximum adsorption capacities of 3.09, 2.59, and 3.47 mmol/L for Ni(II), Cu(II), and Cd (II), respectively. The regeneration of APS/MFC was best accomplished by an alkaline regenerated.

Cellulose-based nanosorbents of magnetic nature were fabricated and employed for the removal of Hg(II), Cu(II), and Ag(I) ions [110]. The process of removal of metallic toxicants was found to be highly efficient, and the uptake capacity was found to be 2, 1.5, 1.2 mmol/g for Hg(II), Cu(II), and Ag(I), respectively. The adsorption was quite fast and the process followed pseudo-second-order kinetics. It was observed that the equilibrium adsorption reached within 5 min. From kinetic considerations, the process was quite fast and exothermic in nature. The adsorbent was successfully regenerated after reaction with acidified thiourea.

The nanoparticles of chitosan were fabricated by carrying out cross-linking of this cationic biopolymer with maleic acid [111] to yield particles of size ranging from 65 to 250 nm. The adsorption potential of the so-prepared nanosorbents for lead ions was investigated following a batch mode and at room temperature. The adsorption was studied as a function of pH, metal on concentration, adsorbent dose, and the experimental conditions were optimized. It was found that in the pH range 3–6, at metal ion concentrations of 10–100 mg/L and adsorbent dose of 1–7.5 g/L, the adsorption was optimum. The authors reported a maximum removal efficiency of 86% in their study.

It was laid down by Salipira et al. [112] that insoluble cyclodextrin polyurethanes have shown promise in removing organic contaminants form water at extremely low concentration, to say at nanograms per liter. It was observed that the carbon nanotubes also possess abnormally high capacity to adsorb organic species like dioxins and polychlorinated dibenzo-furans. However, the high cost of carbon nanotubes with well-defined architecture restricts its commercial and economical use in water remediation. It was also found that when the carbon nanotubes are incorporated into a polymer, the so-produced nanocomposites show extremely well removal potential of organic species like trichloroethylene. This could be another good option to remove organic contaminants from water. The polymers can also be recycled and reused for subsequent adsorption processes.

Fan et al. [113] designed magnetic β-cyclodextrin–chitosan/graphene oxide materials (MCCG) via chemical synthesis and evaluated the efficiency of the prepared materials for the removal of dyes from aqueous solutions. The workers observed that the prepared materials demonstrated outstanding dye removal capacity which was thought to be mainly due to the respective properties of constituent materials such as enormously high surface area of graphene oxide, abundance of amine and hydroxyl groups due to chitosan, fairly higher hydrophobic nature of cyclodextrin, and magnetic properties of iron oxide. The adsorbed dye molecules may easily be recovered by applying external magnetic field so that the iron oxide nanoparticles along with dye molecules get separated. The effect of various factors was studied on the dye removal capacity of the prepared adsorbents, and the adsorption data were applied to various adsorption models to examine their suitability. The materials developed offered notable advantages like high removal capacity, reusability, speedy extractions that enabled materials to apply for the removal of other toxicants species also.

Blends of polysulfone (PSf)/β-cyclodextrin (β-CD)/polyurethane composite membranes were prepared and characterized following a modified phase inversion technique [114]. The purpose of this study was to design nanofiltration membranes for selective removal of cadmium ions from aqueous solutions having concentrations up to 10% and at pH 6.9. It was observed that the prepared blends possessed greater hydrophilic nature and offered higher permeability to water molecules and retained cadmium ions up to 70%. The studies clearly indicated that a judicious blending of β-CD polyurethane with PSf can result in highly efficient and economically viable membranes with excellent metal ions retention capacity.

6 Current Challenges and Future Prospects

Although cationic polymer and biopolymer nonmaterial have greatly contributed to maintaining water quality by removal of variety of toxicants, yet a complete solution is still quite away and only few materials have come to the level of commercial production and large-scale application of these nano- or macroadsorbents. In this way, a major hurdle is to make effective and economic utilization of adsorbents at large scale and for industrial applications besides domestic uses. Another issue of great environmental concern is the environmental protection as new and newer chemical strategies have although been able to produce precise structures with desirable properties, however, the uncontrolled use of chemicals, catalysts, low yield, series of byproducts, stringent experimental conditions has imposed certain limitations to synthesize and use nano and macroadsorbents for removal of toxic metal ions and other toxic contaminants.

Apart from these restrictions and limitations, the field of water remediation has tremendous scope in the years to come. Increasing awareness for environment and pollution has made the public quite sensitive to these issues, and simultaneously the responsibilities of chemists, engineer, and environmentalists have also increased. In the future, demand of specific adsorbents of high-performance and well-designed architectures will certainly be at its ever highest. Designing adsorbents with low cost ever great performance to water remediation has opened up multiple avenues to scientists and produced opportunities to work together to combat the problem of water and other pollutions with the armory of excellent adsorbent molecules.

References

1. Hernandez, N., R.C. Williams, and E.W. Cochran. 2014. The battle for the "green" polymer. Different approaches for biopolymer synthesis: Bioadvantaged vs. bioreplacement. *Organic & Biomolecular Chemistry* 12: 2834–2849. doi:10.1039/C3OB42339E.
2. Mensitieri, G., E. Di Maio, G.G. Buonocore, I. Nedi, M. Oliviero, L. Sansone, and S. Iannace. 2011. Processing and shelf life issues of selected food packaging materials and structures from renewable resources. *Trends in Food Science & Technology* 22 (2–3): 72–80.
3. Jacquel, N., and C.-W. Lo. 2008. Isolation and purification of bacterial poly (3-hydroxyalkanoates). *Biochemical Engineering Journal* 39 (1): 15.
4. Li, P., Y.F. Poon, W. Li, H.-Y. Zhu, S.H. Yeap, Y. Cao, X. Qi, C. Zhou, M. Lamrani, R.W. Beuerman, E.-T. Kang, Y. Mu, C.M. Li, M.W. Chang, S.S. Jan Leong, and M.B. Chan-Park. 2011. A polycationic antimicrobial and biocompatible hydrogel with microbe membrane suctioning ability. *Nature Materials* 10: 149–156.
5. Zhou, J., J. Liu, C.J. Cheng, T.R. Patel, C.E. Weller, J.M. Piepmeier, Z. Jiang, and W.M. Saltzman. 2012. Biodegradable poly (amine-co-ester) terpolymers for targeted gene delivery. *Nature Materials* 11: 82–90.
6. Vroman, I., and L. Tighzert. 2009. Biodegradable polymers. *Materials* 2 (2): 307–344.
7. Ahalya, N., T.V. Ramachandra, and R.D. Kanamadi. 2003. Biosorption of heavy metals. *Journal of Chemistry and Environment* 7 (4): 71–79.
8. Hua, M., S. Zhang, P. Pan, W. Zhang, L. Lv, and Q. Zhang. 2012. Heavy metal removal from water/wastewater by nanosized metal oxides: A review. *Journal of Hazardous Materials* 211: 317–331.
9. Overview and comparison of conventional treatment technologies: Nano-based techniques. In *Proceedings of International Workshop on Nanotechnology, Water and Development, India*, 2006, 10–12.
10. U.S. Environmental Protection Agency Nanotechnology White Paper Prepared for the U.S. Environmental Protection Agency, 2007.
11. Wang, S., C. Wei, W. Wang, Q. Li, and Zhengping Hao. 2012. Synergistic and competitive adsorption of organic dyes on multiwalled carbon nanotubes. *Chemical Engineering Journal* 197: 34–40.
12. Mara, D.D. 2003. *Domestic wastewater treatment in developing countries*, 94–104.

13. Moore, S.K., N.J. Mantua, and E.P. Salathe Jr. 2011. Past trends and future scenarios for environmental conditions favoring the accumulation of paralytic shellfish toxins in Puget Sound shellfish. *Harmful Algae* 10 (5): 521–529.
14. Hutton, R.J., J.J. Landsberg, and B.G. Sutton. 2007. Timing irrigation to suit citrus phenology: a means of reducing water use without compromising fruit yield and quality? *Australian Journal of Experimental Agriculture* 47: 71–80.
15. Soto, M.L., A. Moure, H. Domínguez, and J.C. Parajo. 2011. Recovery, concentration and purification of phenolic compounds by adsorption: A review. *Journal of Food Engineering* 105: 1–27.
16. Milieu Ltd., WRc, RPA URI: http://ec.europa.eu/environment/waste/sludge/pdf/part_iii_report.pdf, 2010, 266. Cited 2014/9/7.
17. Bratby, J. 2006. *Coagulation and flocculation in water and wastewater treatment*, 2nd ed. London, UK: IWA Publishing.
18. Duan, C., N. Zhao, X. Yu, et al. 2013. Chemically modified kapok fiber for fast adsorption of Pb^{2+}, Cd^{2+}, Cu^{2+} from aqueous solution. *Cellulose* 20: 849–860.
19. Fu, F., and Q. Wang. 2011. Removal of heavy metal ions from wastewaters: A review. *Journal of Environmental Management* 92: 407–418.
20. O'Connell D.W., C. Birkinshaw, and T.F. O'Dwyer. 2008. Heavy metal adsorbents prepared from the modification of cellulose: A review. *Bioresource Technology* 99: 6709–6724.
21. Guardia, Pablo, Amilcar Labara, and Xavier Batlle. 2011. Tuning the size, the shape, and the magnetic properties of iron oxide nanoparticles. *Journal of Physical Chemistry* 115 (2): 390–396.
22. Guibal, E. 2004. Interactions of metal ions with chitosan-based sorbents: A review. *Separation and Purification Technology* 38: 43–74.
23. Roberts, G.A.F. 1992. *Chitin chemise*. Houndmills: Macmillan Press Ltd.
24. Johnson, E.L., and Q.P. Peniston. 1982. *Utilization of shellfish waste from chitin and chitosan production*. Westport: Chemistry and Biochemistry of Marine Food Products.
25. Annaduzzaman, M. 2015. Chitosan biopolymer as an adsorbent for drinking water treatment —Investigation on arsenic and uranium. TRITA-LWR LIC, 2015-02, 26 p.
26. Benavente, M. 2008. Adsorption of metallic ions onto chitosan: Equilibrium and kinetic studies. TRITA CHE Report, 44.
27. Sureshkumar, M.K., D. Das, M.B. Mallia, and P.C. Gupta. 2010. Adsorption of uranium from aqueous solution using chitosan-tripolyphosphate (CTPP) beads. *Journal of Hazardous Materials* 184 (1–3): 65–72.
28. Soliman, E.A., S.M. El-Kousy, H.M. Abd-Elbary, and A.R. Abou-zeid. 2013. Low molecular weight chitosan-based schiff bases: Synthesis, characterization and antibacterial activity. *American Journal of Food Technology* 8 (1): 17–30.
29. Blackwell, J., R. Minke, and K.H. Gardner. 1978. Determination of the structures of α- and β-chitins by X-ray diffraction. In *Proceedings of the First International Conference on Chitin/Chitosan*, ed. R.A.A. Muzzarelli, and E.R. Pariser, 108–123. Cambridge, MA: MIT Sea Grant Program, Massachusetts Institute of Technology.
30. Vartiainen, J., and A. Harlin. 2011. Crosslinking as an efficient tool for decreasing moisture sensitivity of biobased nanocomposite films. *Materials Sciences and Applications* 2: 346–354. http://dx.doi.org/10.4236/msa.2011.25045.
31. Crini, G. 2005. Recent developments in polysaccharide-based materials used as adsorbents in wastewater treatment. *Progress in Polymer Science* 30: 38–70.
32. Chunguang, Yua, and Xuena Han. 2015. Adsorbent material used in water treatment—A review. In *2nd International Workshop on Materials Engineering and Computer Sciences IWMECS*.
33. Cai, J., J. Yang, Y. Du, L. Fan, Y. Qiu, J. Li, and J.F. Kennedy. 2006. Purification and characterization of chitin deacetylase from *Scopulariopsis brevicaulis*. *Carbohydrate Polymers* 65: 211–217. doi:10.1016/j.carbpol.2006.01.003.

34. Salehi, E., S.M. Hosseini, S. Ansari, and A. Hamidi. 2016. Surface modification of sulfonated polyvinylchloride cation-exchange membranes by using chitosan polymer containing Fe_3O_4 nanoparticles. *Journal of Solid State Electrochemistry* 20 (2): 371–377.
35. Westergren, Robin. 2006. Arsenic removal using biosorption with Chitosan: Evaluating the extraction and adsorption performance of Chitosan from shrimp shell waste. TRITA IC.
36. Varma, A.J., S.V. Deshpande, and J.F. Kennedy. 2004. Metal complexation by chitosan and its derivatives: A review. *Carbohydrate Polymers* 55: 77–79.
37. Franco, L.D.O., R.D.C.C. Maia, A.L.F. Porto, A.S. Messias, K. Fukushima, and G.M.D. Campos-Takaki. 2004. Heavy metal biosorption by chitin and chitosan isolated from *Cunninghamella elegans* (IFM 46109). *Brazilian Journal of Microbiology* 35 (3): 243–247.
38. Salehia, Ehsan, Parisa Daraeib, and Ahmad Arabi Shamsabadi. 2016. A review on chitosan-based adsorptive membranes. *Carbohydrate Polymers* 152: 419–432.
39. Kamiński, W., and Z. Modrzejewska. 1997. Application of chitosan membranes in separation of heavy metal ions. *Separation Science and Technology* 32 (16): 2659–2668.
40. Liu, C., and R. Bai. 2006. Adsorptive removal of copper ions with highly porous chitosan/cellulose acetate blend hollow fiber membranes. *Journal of Membrane Science* 284 (1): 313–322.
41. Salehi, E., S. Madaeni, L. Rajabi, V. Vatanpour, A. Derakhshan, S. Zinadini, et al. 2012. Novel chitosan/poly (vinyl) alcohol thin adsorptive membranes modified with amino functionalized multi-walled carbon nanotubes for Cu (II) removal from water: Preparation, characterization, adsorption kinetics and thermodynamics. *Separation and Purification Technology* 89: 309–319.
42. Salehi, E., S. Madaeni, L. Rajabi, A. Derakhshan, S. Daraei, and V. Vatanpour. 2013. Static and dynamic adsorption of copper ions on chitosan/polyvinyl alcohol thin adsorptive membranes: Combined effect of polyethylene glycol and aminated multi-walled carbon nanotubes. *Chemical Engineering Journal* 215: 791–801.
43. Elson, C.M., E.M. Bem, and R.G. Acman. 1980. Removal of arsenic from contaminated drinking water by a chitosan/chitin mixture. *Water Research* 14: 1307.
44. Kyzas Gearge, Z., and A. Deliyanni Eleni. 2013. Mercury (II) removal with modified magnetic chitosan adsorbents. *Molecules* 18 (6): 6193–6214.
45. Zhou, Limin, Yiping Wang, Zhirong Liu, and Qunwu Huang. 2009. Characteristics of equilibrium, kinetics studies for adsorption of Hg (II), Cu (II), and Ni (II) ions by thiourea-modified magnetic chitosan microspheres. *Journal of Hazardous Materials* 161: 995–1002.
46. Vartiainen, J., M. Vähä-Nissi, and A. Harlin, Biopolymer films and coatings in packaging applications—A review of recent developments. *Materials Sciences and Applications* 5: 708–718. http://dx.doi.org/10.4236/msa.2014.510072.
47. Rinaudo, M. 2006. Chitin and chitosan: Properties and applications. *Progress in Polymer Science* 31: 603–632. doi:10.1016/j.progpolymsci.2006.06.001.
48. Mohan, Sneha, Oluwatobi S. Oluwafemi, Nandakumar Kalarikkal, Sabu Thomas, and Sandile P. Songca. 2016. Chapter 3, Biopolymers—Application in nanoscience and nanotechnology, nanotechnology. INTECH Publication. http://dx.doi.org/10.5772/62225.
49. Moon, R.J., A. Martini, J. Nairn, J. Simonsen, and J. Youngblood. 2011. Cellulose nanomaterials review: Structure, properties and nanocomposites. *Chemical Society Reviews* 40 (7): 3941–3994.
50. Hebeish, A.A., M.H. El-Rafie, F.A. Abdel-Mohdy, E.S. Abdel-Halim, and H.E. Emam. 2010. Carboxymethyl cellulose for green synthesis and stabilization of silver nanoparticles. *Carbohydrate Polymers* 82 (3): 933–941.
51. Brännvall, E. 2007. Aspect on strength delivery and higher utilisation of strength potential of soft wood kraft pupl fibres. Ph.D. Thesis, KTH Royal Institute of Technology, Stockholm, Sweden.
52. John, M.J., and S. Thomas. 2008. Biofibres and biocomposites. *Carbohydrate Polymers* 71: 343–364.

53. Brett, C.T. 2000. Cellulose microfibrils in plants: Biosynthesis, deposition, and integration into the cell wall. *International Review of Cytology* 199: 161–199. doi:10.1016/S0074-7696 (00)99004-1.
54. Siqueira, G., and J. Bras. 2010. Cellulosic bionanocomposites: A review of preparation, properties and applications. *Polymers* (Basel) 2 (4): 728–765.
55. Liesiene, J., and J. Matulioniene. 2004. Application of water-soluble diethylaminoethylcellulose in oral drug delivery systems. *Reactive and Functional Polymers* 59: 185.
56. Scott, G. (ed.). 2002. *Degradable polymers: Principles and applications*, 2nd ed. Dordrecht: Kluwer Academic.
57. Sirvio, J., A. Honka, H. Liimatainen, J. Niinimaki, and O. Hormi. 2011. Synthesis of highly cationic water-soluble cellulose derivative and its potential as novel biopolymeric flocculation agent. *Carbohydrate Polymers* 86: 266.
58. Liesiene, Jolanta, and Jurgita Kazlauske. 2013. Functionalization of cellulose: Synthesis of water-soluble cationic cellulose derivatives. *Cellulose Chemistry and Technology* 47 (7–8): 515–525.
59. Khan, Fareha Zafar, Masashi Shiotsuki, Fumio Sanda, Yoshiyuki Nishio, and Toshio Masuda. 2008. Synthesis and properties of amino acid esters of hydroxypropyl cellulose. *Journal of Polymer Science Part A: Polymer Chemistry* 46: 2326–2334.
60. Villiers, A. 1891. Sur la fermentation de la fécule par l'action du ferment butyrique. *Comptes Rendus des Seances de l'Academie des Sciences* 112: 536–538.
61. Loftsson, Thorsteinn, Pekka Jarho, Már Másson, and Tomi Järvinen. 2005. Cyclodextrins in drug delivery. *Expert Opinion on Drug Delivery* 2 (2): 335–351.
62. Frömming, K.-H., and J. Szejtli. 1994. *Cyclodextrins in pharmacy*. Dordrecht: Kluwer Academic Publishers.
63. Loftsson, T., and M.E. Brewster. 1996. Pharmaceutical applications of cyclodextrins. 1. Drug solubilization and stabilization. *Journal of Pharmaceutical Sciences* 85: 1017–1025.
64. Klemm, D., B. Philipp, T. Heinze, U. Heinze, and W. Wagenknecht. 1998. *Comprehensive cellulose chemistry*, vol. 1. Weinheim: Wiley-VCH.
65. Ducoroy, L., M. Bacquet, B. Martel, and M. Morcellet. 2008. Removal of heavy metals from aqueous media by cation exchange nonwoven PET coated with β-cyclodextrin-polycarboxylic moieties. *Reactive and Functional Polymers* 68: 594.
66. Norkus, E. 2009. Metal ion complexes with native cyclodextrins. *Journal of Inclusion Phenomena and Macrocyclic Chemistry* 65: 237.
67. Magnúsdóttir, A., M. Másson, and T. Loftsson. 2002. Self association and cyclodextrin solubilization of NSAIDs. *Journal of Inclusion Phenomena and Macrocyclic Chemistry* 44: 213–218.
68. Thiele, C., D. Auerbach, G. Jung, L. Qiong, M. Schneider, and G. Wenz. 2011. Nanoparticles of anionic starch and cationic cyclodextrin derivatives for the targeted delivery of drugs. *Polymer Chemistry* 2: 209–215.
69. Balta, D.K., E. Bagdatli, N. Arsu, N. Ocal, and Y. Yagci. 2008. Chemical incorporation of thioxanthone into β-cyclodextrin and its use in aqueous photopolymerization of methyl methacrylate. *Journal of Photochemistry and Photobiology A: Chemistry* 196: 33.
70. Del Valle, E.M. 1033. Cyclodextrins and their uses: A review. *Process Biochemistry* 2004: 39.
71. Zhao, G.X., H.X. Zhang, Q.H. Fan, X.M. Ren, J.X. Li, and Y.X. Chen. 2010. Sorption of copper (II) onto super-adsorbent of bentonite–polyacrylamide composites. *Journal of Hazardous Materials* 173: 661.
72. Xie, D.M., and W.X. Sun. 2006. Cyclodextrin and polymers supramolecular complexes as biomaterials. *Journal of Materials Science and Engineering* 24: 623.
73. Allabashi, R., M. Arkas, G. Hörmann, and D. Tsiourvas. 2007. Removal of some organic pollutants in water employing ceramic membranes impregnated with cross-linked silylated dendritic and cyclodextrin polymers. *Water Research* 41 (2): 476–486 (Epub 2006).

74. Haung, Z, S. Liu, B. Zhang, L. Xu, and X. Hu. 2012. Equilibrium and kinetic studies on the adsorpiton of Cu (II) from the aqueous phase using a β-cyclodextrin based adsorbent. *Carbohydrate Polymers* 83: 608.
75. Kriz, Z., J. Koca, A. Imberty, A. Charlot, and R. Auzely-Velty. 2003. Investigation of the complexation of (+)-catechin by β-cyclodextrin by a combination of NMR, microcalorimetry and molecular modeling techniques. *Organic & Biomolecular Chemistry* 1: 2590–2595.
76. Ozmen, E.Y., M. Sezgin, A. Yilmaz, and M. Yilmaz. 2008. Synthesis of β-cyclodextrin and starch based polymers for sorption of azo dyes from aqueous solutions. *Bioresource Technology* 99 (3): 526–531.
77. Szejtli, J. 1998. Introduction and general overview of cyclodextrin chemistry. *Chemical Reviews* 98: 1743.
78. Sakairi, N., N. Nishi, and S. Tokura. 1999. Cyclodextrin-linked chitosan: Synthesis and inclusion complexation ability. In: *Polysaccharide applications: Cosmetics and pharmaceuticals*, vol. 737, ed. M A. El-Nokaly, and H.A. Soini, 58–84. ACS Symposium Series.
79. Hu, J., D. Shao, C. Chen, G. Sheng, J. Li, X. Wang, and M. Nagatsu. 2010. Plasma-induced grafting of cyclodextrin onto multiwall carbon nanotube/iron oxides for adsorbent application. *The Journal of Physical Chemistry B* 114: 6779.
80. Rigby, P.G. 1969. Prolongation of survival of tumour-bearing animals by transfer of "immune" RNA with DEAE-dextran. *Nature* 221: 968–969.
81. Kaminski, K., M. Ponka, J. Ciejka, K. Szczubiaka, M. Nowakowska, B. Lorkowska, R. Korbut, and R. Lach. 2011. Cationic derivatives of dextran and hydroxypropylcellulose as novel potential heparin antagonists. *Journal of Medicinal Chemistry* 54: 6586–6596.
82. Misaki, A., M. Torii, T. Sawai, and I.J. Goldstein. 1980. Structure of the dextran of *Leuconostoc mesenteroides* B-1355. *Carbohydrate Research* 84: 273–285.
83. Heinze, T., T. Liebert, B. Heublein, and S. Hornig. 2006. *Functional polymers based on dextran. Advances in polymer science*, vol. 205, 199–291.
84. Hosseinkhani, H., T. Azzam, Y. Tabata, and A.J. Domb. 2004. Dextran-spermine polycation: An efficient nonviral vector for in vitro and in vivo gene transfection. *Gene Therapy* 11: 194–203.
85. Azzam, T.H. Eliyahu, A. Makovitzki, M. Linial, and A.J. Domb. 2004. Hydrophobized dextran-spermine conjugate as potential vector for in vitro gene transfection. *Journal Controlled Release* 96: 309–323.
86. Hucker, G.J., and C.S. Pederson. 1930. Studies on coccaceae XVI. Genus *Leuconostoc*. *New York State Agricultural Experiment Station Technical Bulletin* 167: 3–8.
87. Jeanes, A., W.C. Haynes, C.A. Wilham, J.C. Rankin, E.H. Melvin, M.J. Austin, J.E. Cluskey, B.E. Fisher, H.M. Tsuchiya, and C.E. Rist. 1954. Characterization and classification of dextrans from ninety-six strains of bacteria. *Journal of the American Chemical Society* 76: 5041.
88. Qader, Shah Ali U.L., L. Iqbal, A. Aman, E. Shireen, and A. Azhar. 2006. Production of dextran by newly isolated strains of Leuconostoc mesenteroides PCSIR-4 and PCSIR-9. *Turkish Journal of Biochemistry* 26: 21–26.
89. Lemarchand, C., P. Couvreur, C. Vauthier, D. Costantini, and R. Gref. 2003. Study of emulsion stabilization by graft copolymers using the optical analyzer Turbiscan. *International Journal of Pharmaceutics* 254 (1): 77–82, 0378–5173 (Print).
90. Rotureau, E., M. Leonard, E. Dellacherie, and A. Durand. 2004. Amphiphilic derivatives of dextran: Adsorption at air/water and oil/water interfaces. *Journal of Colloid and Interface Science* 279 (1): 68–77, 0021–9797 (Print).
91. Samal, Sangram Keshari, Martoni Dash, Sandra Van Vlierberghe, David L. Kaplan, Erno Chiellini, Clemens van Blitterswijk, Lorenzo Moroni, and Peter Dubruel. 2012. Cationic polymers and their therapeutic potential. *Chemical Society Reviews* 41 (21): 7147–7194.
92. Naessens, M., A. Cerdobbel, W. Soetaert, and E.J. Vandamme. 2005. Leuconostoc dextransucrase and dextran: Production, properties and applications. *Journal of Chemical Technology & Biotechnology* 80: 845–860.

93. Aumelas, A., A. Serrero, A. Durand, E. Dellacherie, and M. Leonard. 2007. Nanoparticles of hydrophobically modified dextrans as potential drug carrier systems. *Colloids and Surfaces B: Biointerfaces* 59 (1): 74–80, 0927–7765 (Print).
94. Chen, X.G., C.M. Lee, and H.J. Park. 2003. O/W emulsification for the self-aggregation and nanoparticle formation of linoleic acid modified chitosan in the aqueous system. *Journal of Agricultural and Food Chemistry* 51 (10): 3135–3139, 0021–8561 (Print).
95. Kildeevaa, N.R., P.A. Perminova, L.V. Vladmirov, V.V. Novikove, and S.N. Mikhailove. 2009. Mechanism of the reaction of glutaraldehyde with chitosan. *Russian Journal of Bioorganic Chemistry* 35 (3): 360–369.
96. Nasti, A., N.M. Zaki, P.D. Leonardis, S. Ungphaiboon, P. Sansongsak, M.G. Rimoli, and N. Tirelli. 2009. Chitosan/TPP and chitosan/TPP-hyaluronic acid nanoparticles: Systematic optimisation of the preparative process and preliminary biological evaluation. *Pharmaceutical Research* 26: 1918–1930.
97. Lin, A.H., Y.M. Liu, and Q.N. Ping. 2007. Free amino groups on the surface of chitosan nanoparticles and its characteristics. *Yao Xue Xue Bao* 42: 323–328.
98. Cheng, C., J. Deng, B. Lei, A. He, X. Zhang, L. Ma, S. Li, and C. Zhao. 2013. Toward 3D graphene oxide gels based adsorbents for high-efficient water treatment via the promotion of biopolymers. *Journal of Hazardous Materials* 263: 467–478. doi:10.1016/j.jhazmat.2013.09.065.
99. Sankar, M.U., S. Aigal, S.M. Maliyekkal, A. Chaudhary, A. Avula, A. Kumar, K. Chaudhari, and T. Pradeep. 2013. Biopolymer-reinforced synthetic granular nanocomposites for affordable point-of-use water purification. *Proceedings of the National Academy of Sciences* 110: 8459–8464. doi:10.1073/pnas.1220222110.
100. Mahmoudi, M., S. Sant, B. Wang, S. Laurent, and T. Sen. 2011. Superparamagnetic iron oxide nanoparticles (SPIONs): Development, surface modification and applications in chemotherapy. *Advanced Drug Delivery Reviews* 63: 24–46.
101. Zhang, Wei-Xian. 2005. Nanotechnology for Water purification and waste treatment, Frontiers in nanotechnology, US EPA Millennium Lecture Series July 18, Washington D.C.
102. Khatami, Seyed Yavar, and Zahra Hejri. 2015. Optimizing the adsorption conditions of lead from aqueous solutions onto chitosan nano particles. *Journal of Applied Environmental and Biological Science* 4 (11S): 150–159.
103. Zahra Hezri, 2015. Optimizing the adsorption conditions of lead from aqueous solutions on chitosan nanoarticles. *Journal of Applied Environmental and Biological Science* 4 (115):150.
104. Bhavani, A.L., and J. Nisha. 2010. *Dextran—The polysaccharide with versatile uses*. PG & Research Department of Biotechnology, Sengunthar Arts & Science College, Tiruchengode 637205, Namakkal District, Tamilnadu, India.
105. El-Kafrawy, Ahmed Fawzy, Shimaa Mohamed El-Saeed, Reem Kamel Farag, Hend Al-Aidy El-Saied, and Manar El-Sayed Abdel-Raouf. 2017. Adsorbents based on natural polymers for removal of some heavy metals from aqueous solution. *Egyptian Journal of Petroleum* 26: 23–32.
106. Suopajärvi, Terhi. 2015. Functionalized nanocelluloses in wastewater treatment applications. *Acta Universitatis Ouluensis C*, 526.
107. Dubey, Renu, J. Bajpai, and A.K. Bajpai. 2016. Chitosan-alginate nanoparticles (CANPs) as potential nanosorbent for removal of Hg (II) ions. *Environmental Nanotechnology, Monitoring and Management*. http://dx.doi.org/10.1016/j.enmm.2016.06.008.
108. Hokkanen, Sanna, Eveliina Repo, and Mika Sillanpää. 2013. Removal of heavy metals from aqueous solutions by succinic anhydride modified mercerized nanocellulose. *Chemical Engineering Journal* 223: 40–47.
109. Hokkanen, Sanna, Eveliina Repo, Terhi Suopajärvi, Henrikki Liimatainen, Jouko Niinimaa, and Mika Sillanpää. 2014. Adsorption of Ni (II), Cu (II) and Cd (II) from aqueous solutions by amino modified nanostructured microfibrillated cellulose. *Cellulose* 21: 1471–1487.

110. Donia, A.M., A.A. Atia, and F.I. Abouzayed. 2012. Preparation and characterization of nano-magnetic cellulose with fast kinetic properties towards the adsorption of some metal ions. *Chemical Engineering Journal* 191: 22–30.
111. Leyla, Ekhlasi, Younesi Habibollah, Mehraban Zahra, and Bahramifar Nader. 2013. Synthesis and application of chitosan nanoparticles for removal of lead ions from aqueous solutions. *Water and Wastewater* 24 (1): 10–18.
112. Salipira, K.L., B.B. Mamba, R.W. Krause, T.J. Malefetse, and S.H. Durbach. 2008. Cyclodextrin polyurethanes polymerised with carbon nanotubes for the removal of organic pollutants in water. *Water SA* 34 (1): 113–118.
113. Fan, Lulu, Chuannan Luo, Min Sun, Huamin Qiu, and Xiangjun Li. 2013. Synthesis of magnetic β-cyclodextrin–chitosan/graphene oxide as nanoadsorbent and its application in dye adsorption and removal. *Colloids and Surfaces B: Biointerfaces* 103: 601–607.
114. Adams, Feyisayo V., Edward N. Nxumaloa, Rui W.M. Krausea, Eric M.V. Hoek, and Bhekie B. Mamba. 2012. Preparation and characterization of polysulfone/β-cyclodextrin polyurethane composite nanofiltration membranes. *Journal of Membrane Science* 405–406: 291–299.

Alginate-Based Nanosorbents for Water Remediation

A.K. Bajpai, Priyanka Agrawal, Sunil K. Singh and Priyanka Singh

Abstract The present chapter highlights the major concepts of the extraction, preparation, properties, and applications of brown algal mass. In recent years, brown marine weeds have been investigated as the most effective and promising substrates in water treatments. Thus, being motivated by the massive applications of algal masses, the authors have selected the alginate biopolymer as biosorbent and discussed its potential role in biosorption studies. Herein, we have described the nanocomposites of the seaweed alginate and its derivatives in water remediation. The sorption behavior of alginate and derivatives with various toxic heavy metals as well as radioactive elements is summarized, and their relative performance has been examined. The innovation in creation of synthetic derivatives has the potential to empower the next generation of applications for alginates. Further, the global market reports have emphasized on the upcoming continuous research innovation of marine brown seaweed in wastewater treatments in particularly the Asia-Pacific region.

Keywords Alginate · Nanosrobent · Toxic metal ions · Remediation

1 Water Pollution

Water is one of the essential materials required to sustain life on the earth surface, and it covers about 70% of our earth out of which only 0.77% of water is available for drinking purpose, which is also getting contaminated with various types of pollutants. Water pollution is the problem of major concern; India is facing at present as it adversely affects life of millions of peoples every year [1]. The term "water pollution"

A.K. Bajpai (✉) · P. Agrawal P. Singh
Bose Memorial Research Laboratory, Department of Chemistry,
Government Model Science College, Jabalpur 482001, MP, India
e-mail: akbmrl@yahoo.co.in

S.K. Singh
Department of Chemistry, Guru Ghasidas Central University, Billaspur, CG, India

© Springer International Publishing AG 2018
S. Bhardwaj Mishra and A.K. Mishra (eds.), *Bio- and Nanosorbents from Natural Resources*, Springer Series on Polymer and Composite Materials,
https://doi.org/10.1007/978-3-319-68708-7_5

can be defined as "the presence of undesirable substances in the water whose chemical composition or quantity prevents the functioning of natural processes and produces undesirable changes and health effects." Water crisis is rising as a global problem in today's scenario as it directly affects society and the whole nation too. With increase in population growth, the challenge of providing freshwater is increasing day by day, and about 0.78 billion people are currently suffering from various waterborne harmful diseases as they do not have freshwater resources [2].

Water being such a good solvent is never found naturally in a complete pure state, e.g., rainwater contains dissolved gases and other particulates from the atmosphere. Furthermore, existing freshwater resources are gradually becoming polluted and unavailable due to frequently increasing human and industrial activities [3]. Numerous factors which may lead to water pollution occur in the environment. Some of these factors of water pollution came from industrial plants, chemicals, household activities, and other uses of water in the community [4]. It contains two types of impurities, natural and manmade. The natural impurities are not essentially dangerous whereas, pollution caused by human activity includes different type of industrial wastes which contain toxic pollutants. These pollutants can be broadly categorized as- organic and inorganic pollutant.

Organic pollutants which are found in the environment are organic dyes, discharged from textile and other industrial processes into the water. The dyes presently used in industries include methylene blue (MB), Rhodamine B (RhB), methyl orange (MO), Rhodamine 6G (Rh6G) as well as organic chemicals (phenol and toluene). Release of these into lakes or other water sources has become a serious health concern [5], whereas all cationic, i.e., metallic contaminants come under the category of inorganic pollutants which are known as heavy metals.

1.1 Characteristics of Heavy Metal and Its Impact on Environment

The term "heavy metals" refers to any metallic element that has a relatively high density and is toxic or poisonous even at low concentration [6]. These elements are naturally found in the earth's crust, the distribution of heavy metals in the environment is governed by the properties of the metal and influences of environmental factors [7].

Metal constitute an important class of toxic substance which are encountered in numerous occupational and environmental circumstances. Trace amounts of metals are essential to the human body; however, high concentrations can be dangerous leading to a damage of human health, because they are non-biodegradable and can be accumulated in living tissues. Thus, heavy metals are regarded as the most hazardous and toxic metal ions as they easily enter inside our body through several paths and can cause various types of diseases like cancer. Therefore, determination of trace levels of heavy metals is very critical for environmental protection, food and agricultural chemistry and also for monitoring environmental pollution [8, 9].

Among different types of chemical contaminants affecting water resources, heavy metals can be considered as some of the most problematic pollutants due to their non-biodegradable nature and strong toxicity even at low concentrations. The most common heavy metals that are being exposed to human beings are [Hg(II), Pb(II), Cr(III), Cr(VI), Ni(II), Co(II), Cu(II), Cd(II), Ag(I), As(V) and As(III)]. A hazardous concentration of above mentioned metal ion in natural waters is now a worldwide problem.

1.1.1 Sources of Heavy Metal Ion into the Environment

Effluents from textile, leather, tannery, electroplating, galvanizing, pigment and dyes, metallurgical and paint industries contains considerable amounts of toxic metal ions. Lead being one of very important pollutants comes from wastewaters from refinery and are also present in petrol-based wastewaters. Natural sources of chromium is weathered rocks, volcanic exhalations and biogeochemical processes and, also introduced into the environment through electroplating, petroleum refining, leather tanning, wood preserving, textile manufacturing and pulp processing. It exists in both hexavalent and trivalent forms. Mercury is another important toxic and exceedingly bio accumulative heavy metal. Major sources of mercury pollution include anthropogenic activities such as agriculture, municipal wastewater discharges, mining, incineration, and discharges of industrial wastewater [10]. The predominant source of arsenic contamination is through the weathering of geological materials [11]. Nickel is a moderately toxic element as compared to other transition metals. It is a natural element present in earth's crust and are also introduced in the environment through paint and battery processing units [12]. Copper is mainly employed in electric goods industry and brass production from here it enters in water bodies. The toxicity limits along with their toxic effects of some metal ions are mentioned in Table 1.

Heavy metals released into the surface and groundwater has been a major preoccupation for many years because of their increased discharge, acute toxicity, non-biodegradable nature and tendency for bioaccumulation [13]. Low concentration (below 5 mg/L) of heavy metal is difficult to treat economically using chemical precipitation methodologies, hence new cost-effective techniques need to be developed. Some of the treatment technologies prevailed is discussed below.

1.1.2 Various Treatment Technologies

From the above discussion it can be concluded that the removal of metal ions and other pollutants is rising as an important issue. Hence, it is necessary to protect our environment and public health too against the harmful effect of toxicants. Thus, the improvement of water quality is an important area of research globally for scientists. Several treatment technologies are available to reduce the pollutants' concentrations in wastewater including chemical oxidation and reduction, membrane

Table 1 MCL standards and toxic effects of most hazardous heavy metals. Reproduced from Ref. [4], published in an open access journal

Heavy metal	Toxic effects of metal ion	Heavy metal toxicity MCL (mg/L)
Arsenic (As)	Skin manifestations, visceral cancers, vascular disease	0.050
Cadmium (Cd)	Kidney damage, renal disorder, human carcinogen	0.01
Chromium (Cr)	Headache, diarrhea, nausea, vomiting, carcinogenic	0.05
Copper (Cu)	Liver damage, Wilson disease, Insomnia	0.25
Nickel (Ni)	Dermatitis, nausea, chronic asthma, coughing, human carcinogen	0.20
Zinc (Zn)	Depression, lethargy, neurological signs, and increased thirst	0.80
Lead (Pb)	Damage the fetal brain, diseases of kidney, circulatory system, and nervous system	0.006
Mercury (Hg)	Rheumatoid arthritis and disease of kidneys, circulatory and nervous system	0.00003

separation, liquid extraction, ion exchange, electrolytic treatment, electro precipitation, coagulation, flotation, hydroxide and sulfide precipitation, crystallization, ultra filtration, electro dialysis [14].

However, major drawbacks associated with above mentioned processes are incomplete metal removal, high reagent or energy requirements, and generation of toxic sludge or other heavy metal-containing waste products that may sometimes be more toxic than their parent ones. For this reason, additional disposal methods are required. In addition, they are often expensive, especially when the heavy metal concentrations are very low (e.g., 10–100 mg/L) and comes out as ineffective method. Some of the advantages and disadvantages of the various physicochemical methods for water treatment methods are summarized in Table 2 [15].

Among these methods adsorption prevailed, and has emerged out as effective, economical and eco-friendly treatment technique and also considered as suitable for wastewater treatment because of its simplicity and cost-effectiveness. Adsorption is basically a mass transfer process by which a substance is transferred from the liquid phase to the surface of a solid, and becomes bound by physical and/or chemical interactions. Different types of adsorbents are available but the most commonly used one is activated carbon. It has been used for the removal of Cd, Ni, Cr, Cu, etc. [16]. Although activated carbon is the most commonly used adsorbent but it is expensive, so there has to be some other adsorbent particularly which are low cost, eco-friendly and naturally occurring [17]. Adsorbents containing natural polymers are of great interest because of their properties like chemical stability, particular structure, high reactivity and selectivity toward heavy metal ions. In current scenario, biopolymers have attracted more interest due to increasing environmental

Table 2 Advantages and disadvantages of various physicochemical methods. Reproduced with permission from Ref. [15]

Method	Advantages	Disadvantages	Reference No.
Electrodialysis method	Good quality of water is produced because of high separation selectivity	Membranes replacement and the corrosion process	[15]
Chemical precipitation	Low cost, simple operation technique	Produces more toxic sludge hence leads to extra operational cost for sludge disposal	[14]
Ultrafiltration/reverse Osmosis	High separation selectivity	Large amount of wastewater is produced	[15]
Membrane filtration	Small space requirement, low pressure, high separation selectivity	Unable to remove metal ions	[14]
Adsorption	Low cost, easy operating conditions, high metal binding capacities	Low selectivity, production of waste products	–
Adsorption using biopolymers (Biosorption)	Low cost, easily available easy operating conditions, high metal binding capacities	Biosorbents can be reused, biodegradable, shows much better removal efficiency	–

concern, as biopolymers are biodegradable and nontoxic to the environment. Before we discuss about the use of naturally occurring biopolymers for the removal of toxic metal ions we should have some information about biopolymers.

1.2 Biopolymers

Biopolymers are organic polymers that are synthesized by biological organisms. They consist of monomeric units that are bonded into larger formations. Organic polymers such as natural bitumen, straw, and sticky rice have been used in ancient civilizations and can also be classified as biopolymers in a broad sense [18]. Biopolymers have exhibited potential application in different fields such as agriculture, wastewater treatment, food industry, drug delivery. Biopolymers have also attracted a lot of research attention being nontoxic, low cost and eco-friendly. Biopolymers are used in different fields because of their ease in physical properties modification including mechanical properties and degradation behavior that can be readily controlled over a broad range to match the requirements for a particular application [19].

Based on the manufacturing process, biopolymers are classified into two categories viz. synthetic polymers and natural polymers. The synthetic polymers include polyhydroxyalkanoates, polyvinyl alcohol, polylactic acid, polyglycolic acid, and the natural polymers include polysaccharides (e.g., cellulose, chitosan, gums, alginate, agar, starch) and some proteins from plant and animal origin. The natural polymers can further be classified into three categories—polynucleotides (e.g., RNA and DNA), polypeptides (e.g., composed of amino acids), and polysaccharides—among the three types of biopolymers, polysaccharides have been the most commonly applied in various practices [20].

1.2.1 What Are Polysaccharides?

Polysaccharides are polymeric carbohydrate chains composed of monosaccharide units. Polysaccharides are widely found in nature because they are employed in key biological roles, as substances forming skeletal structures, assimilative reserve substances, and water-binding substances [21]. The properties of polysaccharides have led to their worldwide use as thickening agents, stabilizers, sweeteners, and gel-forming agents in the field of food production, agriculture, cosmetics, medical treatment, pharmaceuticals, and also water remediation [22].

There are different types of polysaccharides or biopolymers which were used in the field of water remediation. Using biologically active adsorbents for the removal of toxicants/pollutants from wastewater is known as bioremediation or we can also define biosorption as a science of removal or reduction of pollutants from the environment using biological means.

1.2.2 A Brief Account of Biosorbents or Biopolymers Used for Toxic Metal Ion Removal

Biopolymers are used in different fields because of their ease in physical properties modification including mechanical properties and degradation behavior that can be readily controlled over a broad range to match the requirements for a particular application [9]. The usage of natural materials that are available in large quantities or certain waste from agricultural operations as low-cost adsorbents may be advantageous as they are widely available and are nature friendly. The good adsorption behavior of biopolymers made it applicable in the field of water remediation is mainly attributed to:

- Presence of a large number of functional groups (acetamido, primary amino, and/or hydroxyl groups).
- High hydrophilicity of the polymer.
- High chemical reactivity due to the presence of different functional groups.
- Easy modification of polymeric chain due to the presence of flexible structure.

Different researchers have used different biosorbents in the field of water remediation. Let us discuss about some examples of biosorbents like grafted cellulose were used as an excellent adsorbent for wastewater treatment process [23]. Starch, which is naturally occurring and is obtained from variety of plants, such as maize, rice, wheat, potatoes, and cassava is also used in the field of water remediation. It consists of D-glucose residues linked by α-(1, 4) glycosidic bonds. Starch phosphate carbamate is reported to have high adsorption capacity for Cu(II) ions, and it is pointed out that it may act as a superabsorbent in many applications [24]. Sawdust can be used as a low-cost adsorbent to remove heavy metal ions from water [25]. Natural bamboo sawdust can also be an efficient Cu(II) ion adsorbent [26]. Another example is gelatin which is a denatured collagen which is proteinaceous in nature, generally obtained by the controlled hydrolysis of collagen extracted from animal tissues, such as skin, bovine, and porcine bone, is also employed in water treatment. Chitosan is the partially deacetylated chitin prepared by the alkaline deacetylation. Chitosan among other biosorbents is one of the most promising alternative adsorbents for the recovery of heavy metals from wastewater [27]. All above discussed biopolymers were used in the field of water remediation too, but among all, alginate is the widely used biosorbent as it is an anionic polymer which shows a great efficiency toward metal ions. The adsorption capacity of the algae, i.e., alginate is directly related to the presence of sites such as carboxyl and sulfate on the surface of alginate polymer which makes it efficient in the field of water remediation [28].

Another important functional group is sulfonic which plays a secondary role, except when metal binding occurs at low pH. Hydroxyl groups are also present in polysaccharides which become negatively charged when pH exceeds beyond 10; hence, hydroxyl group also plays a secondary role in metal binding at low pH [29], but at higher pH, it works effectively for the removal of cationic pollutants. Before discussing about its importance in the field of water remediation, let us have a brief knowledge of its occurrence and its properties.

1.3   Sources of Alginate

Alginate, is also known as alginic acid, is an anionic polysaccharide found in cell walls of algae, and for commercial purposes, it is obtained from various species of kelp or brown algae, Laminaria hyperborean, Macrocystis pyrifera, Laminaria digitata, Ascophyllum nodosum, Laminaria japonica, Ecklonia maxima, Lessonianigrescens, Durvilleaantarctica, and Sargassum spp. It is also produced by two bacterial genera, Pseudomonas and Azotobacter [30]. Alginate is natural nontoxic, biodegradable, and biocompatible polysaccharide distributed in brown seaweeds. It is a water-soluble most versatile biopolymers used in a wide range of applications as stabilizing agent. Hydrated alginate develops into gel when exposed to divalent ions [31].

1.3.1 Chemical Structure of Alginate

Alginate polymers are linear unbranched polysaccharides consisting of (1–4)-linked β-D-mannuronic acid (M) and α-L-guluronic acid (G) monomers of varying sequence and are arranged all along the length of the polymeric chain. Generally, they are arranged in three different patterns: consecutive G residues, consecutive M residues, and alternating MG residues as shown in Fig. 1. These polymeric chains of alginate can be separated into fractions by the partial acid hydrolysis. Two fractions are with homopolymer monomers of G and M, while third fraction consists of equal proportions of both monomers. It can be concluded from the above discussion that alginate could be regarded as copolymer composed of homopolymer regions of M and G, interspersed with alternating structure of MG [32] as shown in Fig. 1.

Later on, it was concluded by the researchers that alginates have no regular repeating unit. The arrangement of monomeric unit in the polymer chain is random,

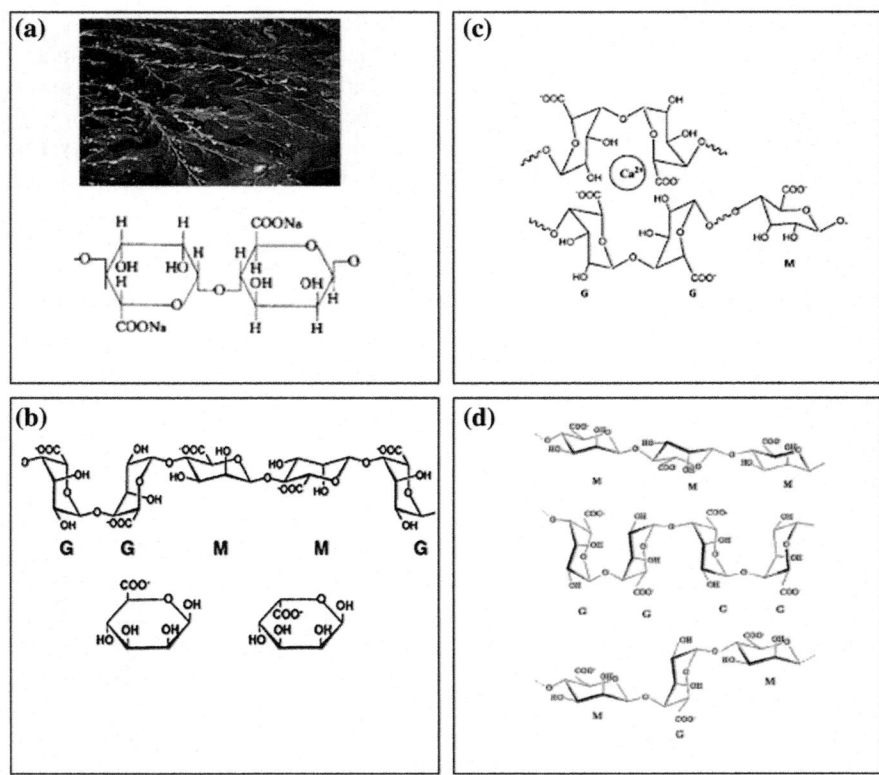

Fig. 1 Pictorial presentation of **a** Natural form of alginate and its molecular structure, **b** Monomeric unit of β-D-mannuronate and α-L-glucuronate, **c** Egg box model, cross-linking structure of alginate with metal ion **d** Polymeric chain of MM, GG, and MG type found in alginate

and therefore, alginates do not contain same repeating units. This property of variation in the polymer provides information about the organism from which the polymer is extracted. The functionality and property of alginate polymer are greatly influenced by the presence of mannuronate and guluronate in different ratio. Properties and functionality of alginate are greatly influenced by the ratio between mannuronate and guluronate monomers in the polymeric chain [33].

1.3.2 Extraction of Alginate

The extraction process can be performed in steps as conversion of insoluble alginate into a soluble form, namely sodium alginate, followed by its precipitation. For the extraction of alginate, the seaweed is broken into pieces and stirred with hot solution of sodium carbonate for about two hours. The solution thus obtained is highly viscous to filter and then diluted with large quantity of water. The diluted extract is left for several hours. The next step is precipitation of the alginate from the above solution, it can be extracted, either as alginic acid or calcium alginate as discussed below.

The first way is to add acid to convert it to alginic acid, which is insoluble in water. After this, alcohol is added to the alginic acid, followed by sodium carbonate which converts the alginic acid into sodium alginate. The sodium alginate does not dissolve in the mixture of alcohol and water, so it can be separated from the mixture as shown in flow diagram Fig. 2.

The second way of recovering the sodium alginate from its solution is to add a calcium salt firstly. This will convert alginate to fibrous calcium alginate which is insoluble in water. The separated calcium alginate is treated with acid to convert it into alginic acid. This fibrous alginic acid is easily separated and placed in alcohol, and sodium carbonate is gradually added to it until all the alginic acid is converted to sodium alginate as shown in flow diagram Fig. 2 [34].

Similar type of method was used by Haug et al. for the extraction of sodium alginate [35], in which first leaching was done with 2% $CaCl_2$ for 1 h and second leaching with 5% HCl for approx 30 min at 30–40 °C, and treated with 40% of formaldehyde for 2 h to cross-link phenolic compounds, the obtained samples were then washed with deionized water and extracted with 5% Na_2CO_3 for 48 h,

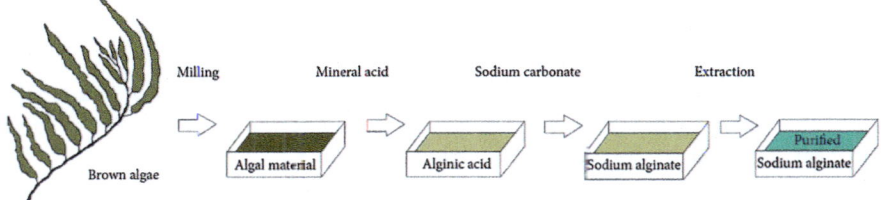

Fig. 2 Extraction procedure of sodium alginate from brown algae

then filtered and precipitated in ethanol as sodium salt. The precipitate was washed with acetone and dried overnight at 60 °C. Thus, the obtained alginate can be used further for the preparation of adsorbents for the removal of toxic metal ion from the water.

Alginate is used as efficient adsorbent in the field of water remediation, but it cannot be used as such, as we all know that its sodium form is soluble in water, and also with change in pH range, it can form a gelatinous mass. Hence, to increase its physical and chemical properties, alginate is cross-linked with polyvalent metal ion and converted into different forms according to our need to increase its removal efficiency. Alginate as adsorbent for the removal of toxic ions from the water can be prepared by different techniques such as emulsification, solvent diffusion method, polyelectrolyte complex (PEC), microemulsion method, ionotropic gelation, desolvation technique, counter ion-induced aggregation are available. Some of the methods have been discussed below for the conversion of alginate in different forms.

1.4 Different Forms of Alginate as Adsorbent and Their Preparation Methods

1.4.1 Emulsion Polymerization Method for the Preparation of Alginate Nanoparticles

In this method for the preparation of nanoparticles, known amount of sodium alginate is dissolved in warm water and stirred over magnetic stirrer till a homogeneous solution is obtained and then same volume of paraffin light oil was added to prepare stable emulsion. To the above emulsion of polyvalent cation is added as cross linker, and then the whole reaction mixture is stirred for about an hr. Now, the gelatinous mass thus obtained is washed several times with toluene and then twice with acetone to remove oil phase [36].

1.4.2 Method for the Preparation of Alginate Beads

Beads of sodium alginate are prepared by dissolving it in required amount of double distilled water. It is left overnight for deaeration. The uniform-sized beads are prepared by adding the solution dropwise in calcium chloride solution with the help of a syringe. The beads so prepared get cross-linked with calcium ions. The stirring was continued for one hour, and the calcium alginate beads were harvested by filtration, washed with distilled water, and air dried overnight [37]. Alginate beads can also prepared by mixing it with other biopolymer such as chitosan, starch, cellulose-like alginate–chitosan (binary beads), chitosan–sodium alginate–cellulose beads (ternary beads).

1.4.3 Method for the Preparation of Alginate Film

Sodium alginate films can be prepared very easily by dissolving known amount of alginate in 100 ml of water under vigorous agitation. For fabricating films, known volume of the above solution was spread into a frame and then dried for 24 h at 50 °C in hot air oven. For preparing Ca-alginate films, the sodium films were soaked in the solution of calcium chloride solution for 3 to h. The films obtained were rinsed with deionized water to remove any excess calcium chloride on the film and dried for 12 h at 50 °C [38].

1.4.4 Preparation of Nanobiocomposites

Nanobiocomposites are the new class of composite materials where nanofillers are incorporated into biopolymer matrix using various techniques, such as (i) in situ polymerization, (ii) solvent intercalation, and (iii) melt intercalation process. Solvent intercalation method is widely used for layered silicates as nanofillers which are to be intercalated in the polymer matrix. It is based on the principle of diffusion in which polymer chain diffuses between the galleries of silicate layers. In this method, solvent is selected in such a way that polymer is soluble in solvent while inorganic nanofillers swell. Polymer is dissolved in solvent, and then, inorganic nanofillers are added in solution with constant stirring. This leads to intercalation of polymer into silicate to form nanobiocomposites [39].

1.5 *Brief History of Alginate in the Field of Bioremediation*

Alka Tiwari and Prerna Kathare worked on the removal of Cu^{2+} ions from aqueous systems using superparamagnetic PVA-Alginate microspheres and concluded that the prepared microspheres came out as an effective adsorbent [40]. The maximum removal of approximately 98% ions was found at pH 4. Tiwari et al. also worked on the removal of organic pollutant phenol and obtained about 80% of removal efficiency from contaminated water [41]. Harikumar and Joseph worked on iron nanoparticles entrapped Ca-alginate bead for the removal of arsenic (III) ion and obtained up to 99.9% removal efficiency at a low adsorbent dose, and in very short time [42]. Singh et al. used iron cross-linked alginate nanoparticles in fixed bed studies for the removal of arsenic ions and found that the nanoparticles worked well in the column and also found effective against bacteriological contaminations [43]. Gomez et al. worked on Na-alginate and Ca-alginate film for the removal of Pb II and obtained 98% of removal efficiency using Ca-alginate film as compared to Na-alginate [44]. Fourest and Lee along with their coworkers studied about removal of heavy metal and Cr(VI) from aqueous solution using alginate and concluded in their research article that alginate is an efficient adsorbent and can be used in the field of water remediation [45, 46]. From the above discussion of research work, it can be concluded that alginate can be used as an effective low-cost biosorbent for the removal of toxic metal from wastewater.

1.6 Alginate Composite as Superabsorbent

The main objectives of the composite absorbent are to enhance the uptake of toxic metal ions, radionuclides, organic and inorganic solutes, bacteria and viruses present in surface water, groundwater, textile wastewater, and industrial wastewater with high hydrophobic and hydrophilic attributes with improved mechanical stability and chemical resistant. The following research papers mention the applications of alginate biopolymer as a superabsorbent composite in the water absorbency that could be effective in developing techniques in water remediation.

A superabsorbent composite (alginate-g-PAMPS/MMT) was prepared by graft copolymerization from alginate, 2-acrylamido-2-methyl-1-propanesulfonic acid (AMPS), and Na+ montmorillonite (MMT) in an inert atmosphere. The introduced montmorillonite formed a loose and porous surface and improved the water absorbency of the alginate-g-PAMPS/MMT superabsorbent composite [47]. In another study, a novel superabsorbent was prepared by graft copolymerization of 2-acrylamido-2-methyl propane sulfonic acid onto alginate in the presence of a cross-linking agent (Fig. 3). The resultant superabsorbent composite had a large degree of water absorbency [48, 49].

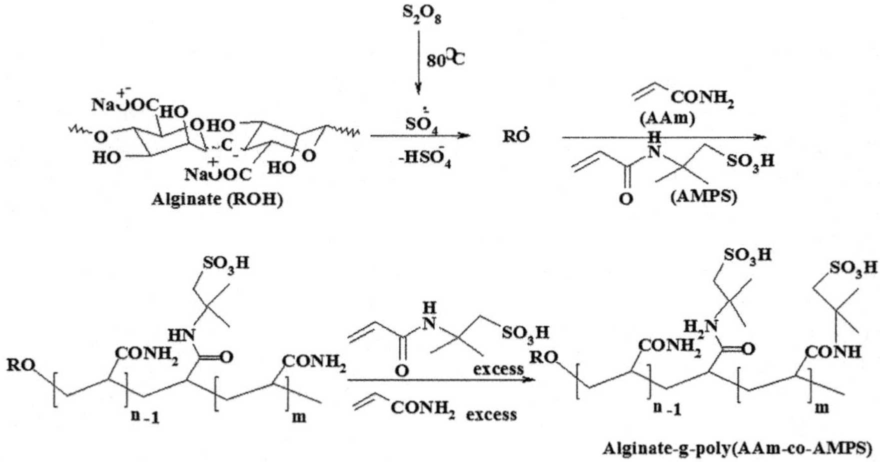

Fig. 3 Proposed mechanistic pathway for synthesis of alginate-based copolymer. Reproduced with permission from Ref. [48], published in an open access journal

1.6.1 Magnetic Alginate Composite as Superabsorbent

In recent years, magnetic-assisted adsorption separation has emerged as a promising technology employed in water treatment. The magnetic sorbents behave similar to or even better than various commercial adsorbents. Magnetic adsorbents can be used to adsorb contaminants from aqueous effluents, and after adsorption, the adsorbents can be separated from the medium by a simple magnetic process. In the literature, iron oxides have been found to be successfully used as composite materials with host materials in fabricating magnetic sorbent [50]. The magnetic sensitivity was achieved by incorporation of magnetic nanoparticles (MNPs) within the alginate gel [51]. The main advantages of using iron oxides as composite materials with host materials are the high porosity, magnetic property, and sometimes good settling property. Since surface functional group reactions are involved in the sorption processes, higher content of surface functional group sites in a sorbent would greatly lead to higher sorption capacity for removal of contaminants [50]. In another study, magnetically separable and stable alginate/Fe_3O_4 composite has been synthesized for the strontium (Sr) removal in complex media, such as seawater and radioactive wastewater. The adsorption experiments for radioactive ^{90}Sr revealed a removal efficiency of 67% in real seawater, demonstrating the reliability of the alginate/Fe_3O_4 composite (Fig. 4 [52]).

1.7 Morphology of Alginate-Based Sorbents

The surface morphology plays an important role in determining the properties of biopolymers in its specific application field. In the following Fig. 5, the scanning electron micrographs (SEM) of calcium alginate microbeads in 50x and 400x magnification and alginate microsphere (SEM, Sirion, FEI, 5 kV) have been depicted to explore the further research applications of brown seaweed in water remediation [53, 54].

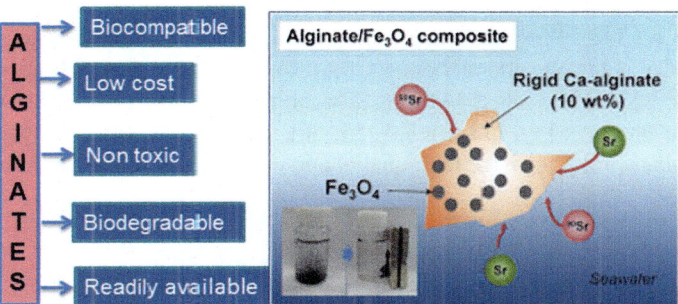

Fig. 4 Characteristics of alginate/Fe_3O_4 composite in strontium (Sr) removal from sea water. Reproduced with permission from Ref. [52]

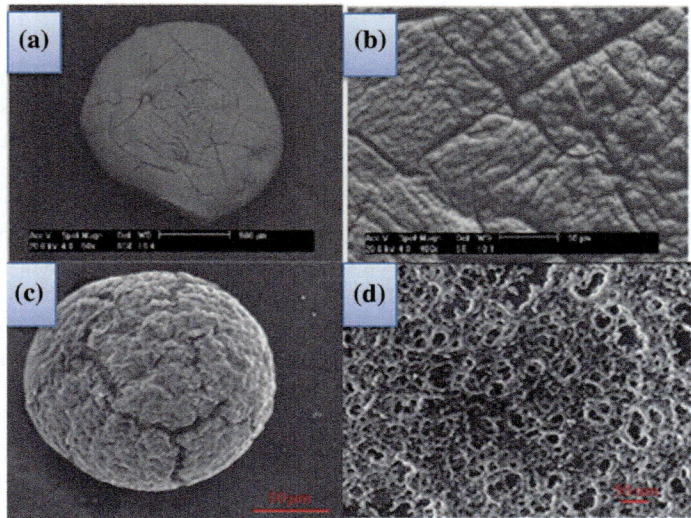

Fig. 5 SEM images of alginate microbeads **a**, **b** in 50x and 400x magnification and alginate microsphere **c**, **d** in (SEM, Sirion, FEI, 5 kV). Reproduced with permission from Ref. [53]

1.8 Market Strategy of Alginate-Based Nanosorbents

The growth in global alginates and derivatives market is due to its multi-functionality properties such as thickening agent, gelling agent, stabilizer, emulsifier, and film-forming property in a wide range of applications.

1.8.1 Key Players: Alginate Market

The major manufacturers of alginates and derivatives include FMC Corporation (USA), Cargill Inc. (USA), E.I. DuPont de Nemours and Company (USA), The Dow Chemical Company (USA), and KIMICA Corporation (Japan), Qingdao Rongde Seaweed Co., Ltd., Prestige Brands, Inc., Incorporated, Qingdao Liyang Seaweed Industrial Co., Ltd., Shandong Jiejing Group Corporation, Prinova Europe Limited, Compañía Española de Algas Marinas S.A., and A2 Trading GmbH, SNAP Natural & Alginate Products Pvt. Ltd. (India) [55, 56]. FMC BioPolymer's advanced manufacturing capabilities provide high quality and consistent grades of alginate [57, 58].

1.8.2 Market Segment: Alginate Market

The segments of the alginates and derivatives market considered for this study include types, applications, and regions (Fig. 6). On the basis of type, the alginates

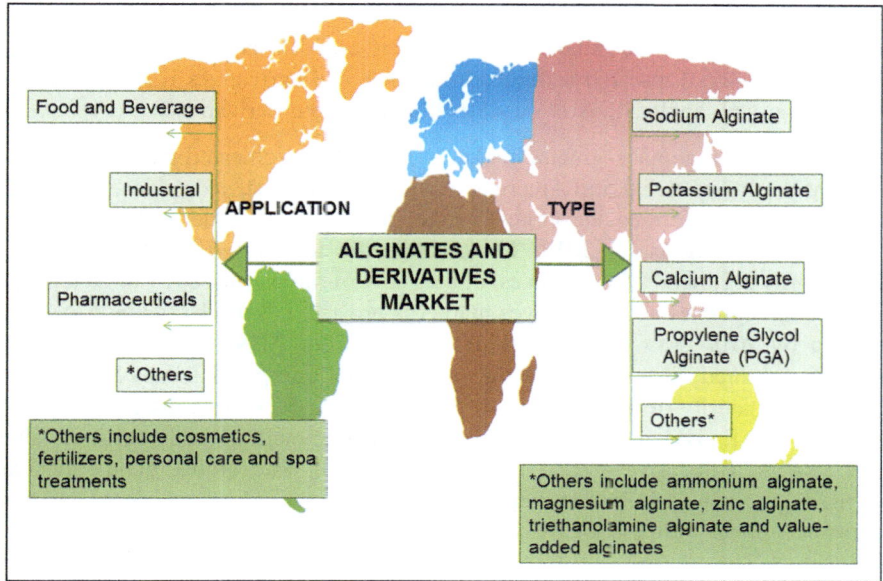

Fig. 6 Global alginate and derivatives market by type and application

and derivatives market is segmented into sodium, calcium, potassium, propylene glycol (PGA), and others. Alginates and derivatives are extensively used in textile printing and other industrial applications. Sodium alginate is the most widely used alginate derivative in foods. The food and pharmaceutical industries have increased their total alginate consumption considerably in the past few years and are expected to grow by the analysis reports from 2014 to 2019, globally, with the increasing demand for food-grade alginates. The growing food and beverage industry in developing economies such as China, Japan, and India has spurred the growth of the alginates industry.

1.8.3 Growth Factors: Alginate Market

The following factors may be considered for the growing global market demand of alginate and its derivatives.

- Growing consumer preference for bakery, confectionery, and dairy products
- Increasing production and extraction of alginates
- Demand for new food variants and ingredients
- Increased import of alginates and ease of availability
- Increasing demand from pharmaceutical and medical industries

- Technical advancement in developing new derivatives alginates
- Cost-effective product innovation in alginate and derivatives
- Growing global research activities in alginate and derivatives
- Finding new applications of developed alginate and derivatives.

On the basis of regional segment, market of alginate is segmented into five regions, namely North America, Latin America, Europe, Middle East, and Africa. The North American region formed the largest market for alginates & derivatives and was valued at 0.3 million, in 2013. The Asia-Pacific region is projected to be the fastest-growing market for alginates, at a CAGR of 5.6% from 2014 to 2019 [56]. The global alginates and derivatives market is projected to reach a value of $409.2 million by 2019 growing at a CAGR of 3.8% from 2014 to 2019. As per the alginate production, major producing regions are Europe and Asia-Pacific and contribute majority of market share in global level market. In terms of value, it is expected that Europe and North America accounted to major share in terms of value and also as a lucrative market in near future [55]. The global trends such as increasing demand for natural ingredients, clean-label products, and products that boost health and wellness are driving the upcoming alginates & derivatives market [59].

1.9 Conclusion and Future Trends

As one looks to the future, the alginate-based materials used in water remediation are likely to evolve considerably. A future progress in this subject will require concentrating efforts on elaboration and development of cost-effective reusability of magnetic biocomposites of alginate. Furthermore, following the novel route to engineer new classes of alginates nanoparticles should also contain advantages in bioremediation. Much study remains to be done on a long way from the easy market availability, a laboratory preparation to effective usability in all its way of synthesis, and the success in this field depends on the effective cooperation of materials scientists, chemists, biologists, government organizations, and in the responses of commercialized alginates products in global market. Further, the global market report emphasizes the upcoming continuous research innovation of marine brown seaweed in wastewater treatment in the Asia-Pacific region due to the growth in emerging economies such as China, Japan, and India.

Furthering our current understating of fundamental properties of alginate and developing new types of metal binding alginate gels may enable future advances in water remediation. The ability to engineer novel sections of alginates with precisely controlled chemical and physical characteristics, unlike the limited repertoire available from natural sources, designed for a specific application could revolutionize the use of these materials. In addition to history of success, we believe that the best is yet to come for alginates in water remediation.

Annexure-I

Legends to Tables

S. No.	Title
Table 1	The MCL standards and toxic effects of most hazardous heavy metals [4]
Table 2	The advantages and disadvantages of various physicochemical methods

Legends to Figures

S. No.	Title
Figure 1	Showing pictorial presentation of (a) Natural form of alginate and its molecular structure (b) Monomeric unit of β-D-mannuronate and α-L-glucuronate (c) Egg box model, cross-linking structure of alginate with metal ion (d) Polymeric chain of MM, GG, and MG type found in alginate
Figure 2	The extraction procedure of sodium alginate from brown algae
Figure 3	Proposed mechanistic pathway for synthesis of alginate-based copolymer [49]
Figure 4	Characteristics of alginate/Fe_3O_4 composite in strontium (Sr) removal from sea water
Figure 5	SEM images of alginate microbeads (a, b) in 50x and 400x magnification and alginate microsphere (c, d) in (SEM, Sirion, FEI, 5 kV) [53, 54]
Figure 6	The global alginate and derivatives market by type and application

References

1. Abhishek, L., A. Karthik R, D.K. Kumar, and G. Sivakumar. 2014. *The International Journal of Innovative Research in Science, Engineering and Technology (IJIRSET)* 3: 17130–17138.
2. Namasivayam. C. 1995. Edited by R.K. Trivedy, Encyclopedia of environmental pollution and control. *Enviro Media* 1: 30–49.
3. Schwarzenbach, R.P., B.I. Escher, K. Fenner, T.B. Hofstetter, U. von Gunter, C.A. Johnson, and B. Wehrli. 2006. *Science* 313: 1072–1077.
4. Tripathi, A., and M.R. Ranjan. 2015. *Journal of Bioremediation & Biodegradation* 6: 1–5.
5. Reife, A., and H.S. Freeman. 1996. *Environmental Chemistry of Dyes and Pigments*. New York: Wiley.
6. Lenntech. 2004. Water Treatment and Air Purification. Published by Lenntech, Rotterdam seweg, Netherlands www.excelwater.com/thp/filters/Water.Purification.htm.
7. Ahmed, R.A., and A.M. Fekry. 2013. *International Journal of Electrochemical Science* 8: 6692–6708.
8. Khlifi, R., and A. Hamza-Chaffai. 2010. *Toxicology and Applied Pharmacology* 1571–1588.
9. Duran, C., A. Gundogdu, V.N. Bulut, M. Soylak, L. Elci, H.B. Senturk, and M. Tufekci. 2007. *Journal of Hazardous Materials* 146: 347–355.

10. Chen, C.W., C.F. Chen, and C.D. Dong. 2012. *International Journal of Geomate* 6: 892–896.
11. Smith, A.H., E.O. Lingas, and R. Mahfuzar. 2000. *Bulletin of the World Health Organization* 78: 1093–1103.
12. Borba, C.E., R. Guirardello, E.A. Silva, M.T. Veit, and C.R.G. Tavares. 2006. *Biochemical Engineering Journal* 30: 184–191.
13. Matschullat, J. 2000. *Science of the Total Environment* 249: 297–312.
14. Huang, X., M. Sillanpä, E.T. Gjessing, S. Peräniemi, and R.D. Vogt. 2011. *River Research and Applications* 27: 113–121.
15. Kurniawan, T.A., et al. 2006. *Chemical Engineering Journal* 118: 83–98.
16. Kwon, J.S., S.T. Yun, J.H. Lee, S.O. Kim, and H.Y. Jo. 2010. *Journal of Hazardous Materials* 174: 307–313.
17. Gottipati, R., and S. Mishra Susmita. 2012. *Research Journal of Chemical Sciences* 2: 40–48.
18. FuWei, Y., Z. BingJian, P. ChangChu, and Z. YuYao. 2009. *Environmental Technological Sciences* 52: 1641–1647.
19. Rhim, J.W., H.M. Park, and C.S. Ha. 2013. *Progress in Polymer Science* 38: 1629–1652.
20. Kalia, S., and L. Averous. 2011. *Biopolymers: Biomedical and environmental applications*. Hoboken, NJ, USA: Wiley.
21. Saha, D., and S. Bhattacharya. 2010. *Journal of Food Science and Technology* 47: 587–597.
22. Lorenzo, G., N. Zaritzky, and A. Califano. 2012. *Food Hydrocolloids* 30: 672–680.
23. Wen, S.S.Y., M.L. Rahman, S.E. Arshad, N.L. Surugau, and B. Musta. 2013. *Journal of Applied Polymer Science* 124: 4443–4453.
24. Guo, L., S.F. Zhang, B.Z. Ju, and J.Z. Yang. 2006. *Carbohydrate Polymers* 63 (4): 487–492.
25. Ahmed, S.A. 2011. *Carbohydrate Polymers* 83 (1470): 1478.
26. Zhao, G., X. Wu, X. Tan, and X. Wang. 2011. *The Open Colloid Science Journal* 4: 19–31.
27. Kavianinia, I., P.G. Plieger, N.G. Kandile, and D.R.K. Harding. 2012. *Carbohydrate Polymers* 87: 881–893.
28. Davis, T.A., B. Volesky, and A. Mucci. 2003. *Water Research* 37: 4311–4330.
29. Gupta, V.K., and A. Rastogi. 2008. *Colloids Surf B Biointerfaces* 64:170–178.
30. Robitzer, M., and F. Quignard. 2011. *International Journal of Chemistry* 65: 81–84.
31. Norton, I.T., W. Frith, and S. Ablett. 2006. *Food Hydrocolloids* 20: 229–239.
32. Frampton, J.P., M.R. Hynd, M.L. Shuler and W. Shain. 2011. *Biomedical Marterials* 6: 1–18.
33. Vauchel, P., K. Leroux, R. Kaas, A. Arhaliass, R. Baron, and J. Legrand. 2009. *Bioresource Technology* 100: 1291–1296.
34. Pawar, S.N., and J.E. Kevin. 2012. *Biomaterials* 954: 3729–3305.
35. Haug, A., S. Melsom, and S. Omang. 1974. *Environmental Pollution* 7: 179–192.
36. Mandal, S.S., S.S. Kumar, B. Krishnamoorthy, and S.K. Basu. 2010. *Brazilian Journal of Pharmaceutical Sciences* 46: 785–793.
37. Vijayalakshmi, K., T. Gomathi, and P.N. Sudha. 2014. *Der Pharmacia Lettre* 6: 65–77.
38. Arica, M., C. Arpa, A. Ergene, G. Bayramoglu, and O. Genç. 2003. *Carbohydrate Polymers* 52: 167–174.
39. ManguaL, J.O., S. Li, H.J. Ploehn, A.D. Ebner, and J.A. Ritter. 2010. *Journal of Magnetism and Magnetic Materials* 322: 3094–3100.
40. Tiwari, A., and P. Kathane. 2013. *International Research Journal of Environmental Sciences* 2: 44–53.
41. Tiwari, A., A. Soni, and A.K. Bajpai. 2012. Synthesis and reactivity in inorganic, metal, organic and nanometal chemistry 42: 1158–1166.
42. Harikumar, P.S., and L. Joseph. 2012. *International Journal of Plant, Animal and Environmental Sciences* 2: 159–166.
43. Singh, P., S.K. Singh, J. Bajpai, A.K. Bajpai, and R.B. Shrivastava. 2014. *Journal of Materials Research and Technology* 3: 3195–3202.
44. Gomez, A., K. Wrobel, S. Kazunori, and T.W. Tzu. 2012. *International Congress on Informatics, Environment, Energy and Applications-IEEA*, vol. 38, IACSIT Press, Singapore.
45. Fourest, E., and B. Volesky. 1995. *Environmental Science and Technology* 30: 277–282.

46. Lee, I., C.G. Leea, J.A. Parka, J.K. Kanga, S.Y. Yoon, and S.E. Kim. 2013. *Desalination and Water Treatment* 51: 3438–3444.
47. Yadav, M., and K.Y. Rhee. 2012. *Carbohydrate Polymers* 90 (1): 165–173.
48. Sadeghi, M., M. Esmat, F. Shafiei, L. Mansouri, and H. Shasava. 2014. *Oriental Journal of Chemistry* 30 (1): 247–253.
49. Bajpai, A.K., and Giri, A. 2003. *Carbohydrate polymers* 53: 271–278.
50. Agrawal, P., and A.K. Bajpai. 2011. *Toxicological and Environmental Chemistry* 93 (7): 1277–1297.
51. Degen, P., S. Leick, F. Siedentiedel, and H. Rehage. 2012. *Colloid and Polymer Science* 290 (2): 97–106.
52. Hong, H.J., H.S. Jeong, B.G. Kim, J. Hong, et al. 2016. *Chemosphere* 165: 231–238.
53. Lezeharia, M., J.-P. Baslya, M., O. Baudua. 2010. *Bouras Colloids and Surfaces A: Physicochemical Engineering Aspects* 366: 88–94.
54. Chen, W., J.H. Kim, D. Zhang, K.H. Lee, et al. 2013. *Journal of the Royal Society, Interface* 10 (88): 1–10.
55. Alginate Market: Global Industry Analysis and Forecast 2016–2024 http://www.persistencemarketresearch.com/market-research/alginate-market.asp.
56. Alginates & Derivatives Market—Global Industry Analysis, Size, Share, Growth, Trends and Forecast 2016–2024 www.marketsandmarkets.com.
57. Alginates & Derivatives Market—Global Trends & Forecast to 2019 http://www.fmcbiopolymer.com/Food/Ingredients/AlginatesPGA/Introduction.aspx.
58. Commercial Seaweed Market Analysis By Product (Brown Seaweed, Red Seaweed, Green Seaweed), By Form (Liquid, Powdered, Flakes), By Application (Agriculture, Animal Feed, Human Consumption) And Segment Forecasts To 2024 http://www.grandviewresearch.com/industry-analysis/commercial-seaweed-market.
59. Alginates & Derivatives Market worth $409.2 Million by 2019 http://www.marketsandmarkets.com/PressReleases/alginates-derivatives.asp.

Chitosan-Based Natural Biosorbents: Novel Search for Water and Wastewater Desalination and Heavy Metal Detoxification

Ankita Dhillon and Dinesh Kumar

Abstract Due to worldwide increasing population, the people may face increasing freshwater scarcity globally. Further, groundwater contamination by health hazardous toxicants like fluoride, arsenic, nitrate, phosphate, mercury, heavy metals, and many more is likewise increasing. Therefore, a number of low-cost and fast water and wastewater treatment methods are being utilized at the present time. A great amount of research has been accomplished to discover efficient water purification methods at lower cost as well as minimum utilization of chemicals and negligible impact on the environment. Hence utilization of chitosan-based natural biosorbents in water and wastewater treatment is a fast-emergent field of interest to environmental consultants and public interest groups. This chapter covers an assortment of chitosan-based natural biosorbents in the field of water remediation. Herein, we include the synthesis of new materials, modification of natural chitosan materials and their utilization for water purification.

Keywords Chitosan · Detoxification · Desalination · Wastewater

1 Introduction

Chelating resins have been successfully utilized for both the removal of heavy metal ions from industrial wastewater and water desalination [1]. Various chelating resins having different functional moieties are utilized for detoxification of wastewater and drinking water. They can be obtained from naturally occurring polymeric matrix or synthetic polymeric matrix. Deacetylation of chitin in alkaline medium leads to the development of chitosan which is a natural polymer [2] and it

---

A. Dhillon
Department of Chemistry, Banasthali University, Tonk 304022, Rajasthan, India

D. Kumar (✉)
School of Chemical Sciences, Central University of Gujarat, Gandhinagar 382030, Gujarat, India
e-mail: dinesh.kumar@cug.ac.in

© Springer International Publishing AG 2018
S. Bhardwaj Mishra and A.K. Mishra (eds.), *Bio- and Nanosorbents from Natural Resources*, Springer Series on Polymer and Composite Materials,
https://doi.org/10.1007/978-3-319-68708-7_6

is mainly made of β (1→4) linked 2-amino-2-deoxy-D-glucopyranose units and residual 2-acetamido-2-deoxy-D-glucopyranose unit. Chitosan has various applications including desalination of water and heavy metal ion detoxification due to its harmless nature and easy biodegradability [3, 4]. It is an excellent natural biosorbent for metal ion detoxification in near-neutral samples due to the availability of many NHF functional groups [5–7].

Natural chitosan has been both physically and chemically modified using various methods to achieve high adsorption capacity. Naturally, chitosan is found as flakes or powder that has narrow applicability mainly for column treatments caused by swelling, low mechanical strength, etc. Chitosan beads have been modified through various routes to conquer these shortcomings [8]. Cross-linking of chitosan with glutaraldehyde or epichlorohydrin is the example of chitosan chemical modifications to prevent their acid liability and to advance metal sorption capacity and selectivity [9]. Although cross-linking can decrease the adsorption capacity due to the reduction in the number of free amino groups, yet it ensures constancy of the polymer in the reaction medium. The adsorption capacity of chitosan is varied with porosity, crystallinity, percent deacetylation and the amount of free amino group [10]. Chitosan can be designed in different shapes, membranes, microspheres, gel beads, films, nanoparticles and nanofibres [11–13]. It provides surface area/mass ratio demonstrating maximum adsorption capacity with minimum column clogging and friction loss [14]. Further, chitin and chitosan have been modified using various synthetic materials, for example polymers that support easy handling and versatility. However, the sorption capacity reduced due to adsorption sites blockage [15]. Different chitosan-based natural biosorbents have been developed with high adsorption capacity and utilized for water and wastewater desalination and heavy metal detoxification [16–30]. This chapter provides valuable information on water and wastewater desalination and heavy metal detoxification by means of chitosan-based natural sorbents and to illustrate the adsorption capacity under various laboratory conditions.

2 Characteristics of Natural Chitosan

Partial or total N-deacetylation of chitin poly-$\beta(1\rightarrow4)$-2-acetamide-2-deoxy-D-glucopyranose leads to the development of chitosan, poly-$\beta(1\rightarrow4)$-2-amino-2-deoxy-D-glucopyranose polysaccharide (Fig. 1). Chitosan is soluble in water even at high degree of deacetylation (>40%) [31]. However, it is usually insoluble in water at pH ~7.0 and ordinary organic solvents (e.g. dimethyl sulfoxide, dimethylformamide, organic alcohols, pyridine) even though the presence of hydrophilic functional groups in the polymer backbone. The insolubility in both aqueous and organic solvents is due to the crystalline structure of chitosan, which is because of high intramolecular and intermolecular hydrogen bonding between the chains and sheets, correspondingly [32]. At 40–50% degree of deacetylation,

Fig. 1 Structure of chitin and chitosan

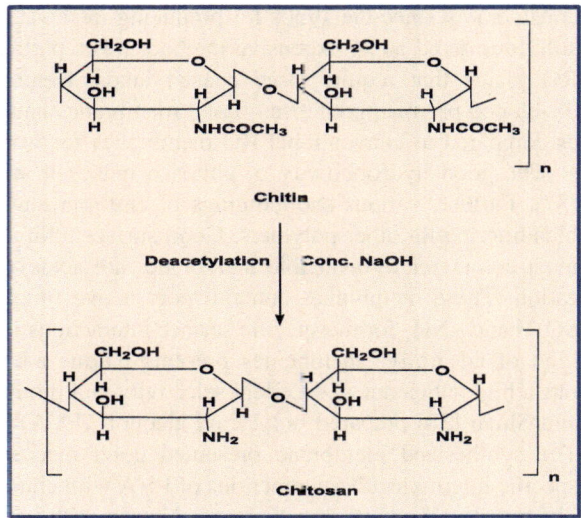

chitosan becomes soluble in acidic media because of protonation of the NH_2 groups on the C2 position of the D-glucosamine unit [33].

Although, the surface area of chitosan is very low, ranging between 2 and 30 m^2/g compared to most commercial activated carbons range between 800 and 1500 m^2/g [34]. The metal adsorption capacity of chitosan has been improved by grafting new functional groups on the chitosan backbone. Further, the sorption selectivity, density of sorption sites and pH range for metal sorption have been improved by incorporating new functional groups in the chitosan backbone [35].

Cross-linking agents have functional groups separated by various molecules that can be structured in different forms (branched chains, rings, straight chains). Di/poly functional reagents can be utilized for partial cross-linking that facilitates the utilization of chitosan in metal ion adsorption in acidic medium. On the one hand, the adsorption capacity was reduced with cross-linking because of decreased number of reactive sites on the chitosan polymer. On the other hand, improvement in the adsorption capacity can be observed depending on the functional groups in the cross-linking agent [36].

3 Chitosan-Based Sorbents for Desalination

Freshwater scarcity has become an increasingly global issue. Unavailability of freshwater leads to understand the primary necessity of saltwater desalination as a long-standing water supply alternative. Desalination offers the smart opportunity of growing the natural water cycle by adding sea water and brackish water. Desalination using reverse osmosis (RO) techniques has been advanced considerably since the 1950s for producing freshwater even though, desalination is still considered as an expensive method. More particularly, it is costly to develop RO plants that require large coastal lands. Hence, various studies have been developed promising chitosan-based membranes having a much high permeability as compared to conventional RO membranes for water desalination.

The good hydrophilicity of chitosan makes it an efficient membrane material [37]. Further, various shortcomings of chitosan have been overcome recently by blending it with other polymers. Good surface affinity and adsorptive properties of most commercially available membranes are achieved using their surface modification. These membranes contain more active functional groups like –COOH, –SO_3H and –NH_2 for the specific surface interactions to the targeted molecules. Such kind of adsorptive membranes presents unique advantages like rapid separation rates, high efficiencies, excellent selectivity and minimum power necessities. Juang and Shiau [38] prepared poly(vinyl alcohol) (PVA) blended chitosan membranes. The synthesized membrane presented good mechanical strength as a result of specific intermolecular interactions of PVA with chitosan. Additionally, the authors illustrated various combinations of chitosan with high strength polymers as effective alternatives for practical desalination applications. Similarly, EL-Gendi et al. [39] utilized polyamide-6 as the polymer matrix due to its good chemical stability and high mechanical strength. On the other hand, chitosan has been utilized as the functional polymer to provide adsorptive properties of the membrane. The increase in chitosan concentration increased both membrane permeate flux and salt rejection.

The high-flux membranes utilize less energy for the production of a calculated amount of water due to low-pressure requirements. Therefore, Raval et al. [40] carried out surface modification of thin-film composite reverse osmosis (TFC RO) membrane using chitosan for advanced water treatment. For that the membrane was treated with sodium hypochlorite solution followed by chitosan treatment. The availability of free –OH groups and modification in polyamide –CO– group led to supramolecular assemblage of chitosan over polyamide. The developed composite membrane showed a low contact angle and high hydrophilicity. Up to 2.5 times increase in flux was observed with solute rejection of chitosan-treated membrane. A similar kind of behaviour was observed with increased sodium hypochlorite exposure, but a simultaneous reduction in solute rejection was also observed. Chitosan-modified membrane showed higher transmembrane flux per °C rise in feed water temperature than a virgin TFC RO membrane. Such high-temperature sensitivity presented it an efficient applicant for solar-powered reverse osmosis. The lower thermal energy increased the feed water temperature, which in turn provided marked benefits in transmembrane flux. The presence of multivalent ions in natural sea water or river water reduces the membrane performance. Therefore, Li et al. [41] developed membranes modified by chitosan/polyaniline composites that showed selectivity for monovalent ions. They have been carried out electrodeposition with chitosan/aniline polymer to utilize it as a modification material for the coating of commercial anion exchange membrane. Grafting of chitosan with polyaniline was achieved by co-polymerization. The results showed that optimum modification was achieved using electrodeposition time of 4 h and aniline ratio as

0.4. The conductivity of membrane rose with the aniline ratio and selectivity firstly increases with the aniline ratio up to a certain point and further reduces. The modified membrane showed high selectivity in sea water with minimum Ca(II) and Mg(II) escape.

Chitosan membranes with appropriate support are more durable in filtration applications as compared to chitosan membrane only. Therefore, Padaki et al. [42] prepared polypropylene fibre supported chitosan nanofiltration membrane by dissolving chitosan in 2% acetic acid. Two TGA peaks were observed at ~90 and ~170 °C in a thermal study of developed membrane due to chitosan and supporting polypropylene membrane, respectively. The lower swelling ratio of membrane was achieved at pH 7.0, 9.0 and 11.0 in contrast to pH 5.0. Largest percentage rejection of prepared membrane was noted in acidic pH and smallest in basic pH. Therefore, the developed chitosan membrane demonstrated high performance in acidic pH media in comparison to basic media. Kumar et al. [43] reported a modified procedure to prepare polysulphone–chitosan (PSf–CS) ultrafiltration membrane due to the insolubility of chitosan in organic solvents. Firstly, two different compositions of polysulfone in N-methylpyrrolidone (NMP) and chitosan in 1% acetic acid were prepared. Then their solutions were blended to develop PSf–CS membranes using diffusion-induced phase-separation method. The water uptake and contact angle measurements demonstrated that PSf–CS membrane had better hydrophilicity than a PSf ultrafiltration membrane. Further, PSf–CS membrane had good antifouling property than PSf ultrafiltration membrane and these properties of PSf–CS membranes were found to increase with chitosan composition. The maximum membrane flux was obtained at the lowest pH as a result of protonation–NH_2 groups of chitosan.

In a study, Ayoub et al. [44] utilized hemicellulose with chitosan for the desalination and heavy metals removal characteristics. The authors carried out alkali treatment of pine wood, switchgrass and coastal Bermuda grass at 75 °C for the extraction hemicelluloses that showed ~450 number average degree of polymerization. This was subsequently utilized for the production of hemicellulose-DTPA-chitosan which involved grafting with penetic acid (diethylene triamine pentaacetic acid, DTPA) and cross-linking with chitosan. The biosorbent demonstrated highest salt removal capacity of ~0.30 g/g. The batch sorption experiments implied the involvement of second-order rate kinetics of the adsorption process. The biosorbent showed 2.90, 0.95 and 1.37 mg/g of Pb(II), Cu(II) and Ni(II) ions uptake, correspondingly at pH 5, and the initial concentrations of metal ions were 5000 ppb. Chen et al. [45] developed graphene oxide–chitosan composite (GO–CS) hydrogels, where 2D GO sheets were cross-linked by CS chains to develop 3D network. The synthesized GO–CS hydrogels presented good mechanical strength due to strong electrostatic and hydrogen bonds between GO and CS. The high porosity of the hydrogels assisted the adsorbates diffusion, thereby enhanced the GO–CS composite adsorption capacity. The GO–CS hydrogel showed high adsorption capacities (>300 mg/g) for both cationic and anionic dyes. The results demonstrated increased adsorption capacity of the hydrogel composite for methylene blue with increased GO content. On the other hand, the GO–CS hydrogel showed the increased adsorption capacity for Eosin Y with increased chitosan content. The mechanism of dye uptake by GO–CS hydrogel involved the affinity between the hydrogel and dyes originated forms an electrostatic attraction. The GO–CS hydrogel showed 70 and 90 mg/g adsorption capacities for Cu(II) and Pb(II) ions, correspondingly. Therefore, GO–CS hydrogel showed potential ability towards column treatment for water purification by filtration. Another group of researchers have investigated polysaccharide hydrogels for softening saline seawater and desalination of reverse osmosis brine [46]. The researchers carried out grafting of acrylamide onto alginate on chitosan using microwave (MW) and ultraviolet (UV) irradiation techniques to develop alginate (Alg-UV and Alg-MW) and chitosan (Ch-UV and Ch-MW) hydrogels. Hydrogel products showed maximum swelling ratios of 168 and 173 g/g for Alg-UV- and Ch-MW-grafted acrylamide in distilled water, correspondingly. The pre-swollen hydrogel demonstrated higher selectivity towards calcium adsorption in sea water and magnesium adsorption in brine as compared to dry hydrogels at the similar adsorbate volume. Both dry and pre-swollen hydrogels showed adsorption capacities of 54 and 34 mg/g for calcium with Alg-MW and UV-prepared alginate and chitosan hydrogels, correspondingly. Further, dry alginate and chitosan hydrogels prepared by MW technique showed maximum magnesium adsorption capacities of 280 and 316 mg/g, correspondingly. The synthesized hydrogels showed higher magnesium adsorption capacities as compared to commercial resins.

4 Chitosan-Based Sorbents for Heavy Metal Ion Detoxification

Heavy metal ion pollution is the major problem worldwide as these ions are toxic to human health. Various industrial activities resulted in the discharge of toxic heavy metal ions into the environment producing the major environmental contamination. A lot of research in the past three years has been done on the utilization of chitosan-based adsorbents for heavy metal ion detoxification from water and wastewaters. Chitosan has a high adsorption affinity for heavy metal ions as a result of high hydrophilicity because of numerous hydroxyl groups of glucose moieties, the existence of abundant functional groups presenting good chemical activity, and the presence of polymer chain having a flexible composition (Fig. 2).

In a study, chitosan was modified by 3,4-dimethoxybenzaldehyde for the detection and removal of Cd(II) from waters [47]. The formation of 3,4-dimethoxybenzaldehyde chitosan derivative (Chi/DMB) was established by FTIR studies. The presence of two new absorption bands in Chi/DMB, one at 1655 cm^{-1} and another at 1562 cm^{-1} showed the presence of an imine bond (C=N) and an ethylenic bond (C=C), respectively. Whereas, the featured free aldehyde group (1720 cm^{-1}) band was disappeared in the product. The pH effect experiments established an increase in adsorption capacity with the solution pH (1.0–9.0).

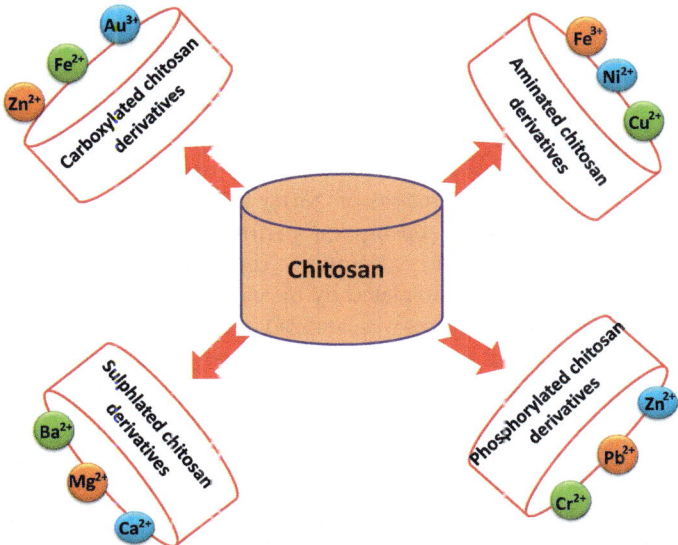

Fig. 2 Heavy metal ions adsorption using different derivatives of chitosan

On the other hand, at pH >7.0, precipitation of cadmium occurred from solution. Therefore, the higher adsorption capacity beyond pH 7.0 can be attributed to the presence of both adsorption and precipitation phenomena on the adsorbent surface. Therefore, pH 6.5 was selected for maximum adsorption, when the precipitation process was neglected. The adsorbent displayed a maximum adsorption capacity of 217.4 mg/g as calculated from the Langmuir isotherm model.

To improve functionality and robustness of adsorbent, Chen et al. [48] carried out the removal of the U(VI) from wastewaters by chitosan-modified multiwalled carbon nanotubes (MWCNT–CS). The developed adsorbent showed maximum adsorption capacity of 71 mg/g at pH 7. In another study, xanthate-modified magnetic cross-linked chitosan (XMCS) synthesized by Chen and Wang for the removal of Co(II) from aqueous systems [49]. The utilization of xanthate introduced thiol groups on chitosan that resulted in an enhanced sorption capacity towards Co (II). The synthesis of the XMCS adsorbent was achieved by xanthate-modification using cross-linked magnetic chitosan with NaOH and CS_2. Stirring and washing of the obtained derivative were carried out at room temperature for 24 h, and the final derivative was dried at 70 °C. The developed adsorbent demonstrated maximum adsorption capacity of 18.5 mg/g.

The magnetically modified chitosan adsorbents have the property of easy separation and therefore they can be easily regenerated and reused. Therefore, Debnath et al. attempted Cr(VI) removal using magnetically modified graphene oxide–chitosan composite [50]. The synthesis of graphene oxide was carried out based on the Hummers method [51]. The maximum adsorption capacity was ∼75 mg/g at 25 °C and at pH 3.0.

Similarly, Elwakeel et al. developed a magnetically modified chitosan resin [52]. Initially, the chitosan was cross-linked with glutaraldehyde in the presence of magnetite. Finally, the chemical modification of cross-linked chitosan was achieved by treatment with tetraethylenepentamine. The developed resin selectively removed $UO_2(II)$ with an adsorption capacity of 1.8 mmol/g at pH 4.0 at room temperature. Chemical modification of raw chitosan achieved with histidine (HIS–ECH–CB) by Eser et al. for the selective removal of Ni(II) [53]. Further, the integration of chelating groups increased the adsorption performance. The resultant-modified chitosan material showed an adsorption capacity of 55.6 mg/g. Firstly, cross-linking of chitosan was completed by means of epichlorohydrin, and subsequent immobilization of histidine was realized after Na_2CO_3 washing. For that obtained beads were placed in a receiver having 10% (w/v) histidine solution and subsequently washed with Na_2CO_3 solution. Finally, the mixing of the materials was done by stirring at 60 °C for 24 h. The beads were then repeatedly washed to remove excess non-immobilized histidine.

A series of modified chitosan beads (CB) were prepared to remove Fe(III) from aqueous solutions [54]. Low adsorption capacities of 7.042, 9.346 and 14.286 mg/g were found for the modified forms of chitosan beads (protonated (PCB), carboxylated (CCB) and grafted CB (GCB), respectively. Alternatively, another grafted chitosan derivative was prepared for the removal of many metal ions like Cu (II), Co(II), Zn(II), Hg(II), Pb(II) from water [55]. The chitosan was modified using 4,4′-diformyl-α-ω-diphenoxy-ethane (A1) or 4,4′-diformyl-2,2′-dimethoxy-α-ω-diphenoxy-ethane (A2). The adsorbent having A2 grafting showed the maximum capacity for Hg(II) followed by Pb(II), Cu(II), Zn(II) and Co(II) ions at solution pH 5.0.

Li et al. developed ethylenediamine-modified yeast biomass by coating with magnetic chitosan microparticles (EYMC) [56]. The employment of magnetic chitosan facilitated successful regeneration of the adsorbent with negligible loss of adsorption capacity. The adsorbent presented interesting morphological characteristics on the surface of the developed EYMC material. SEM images displayed numerous, small bumps on the surface of the adsorbent with large number of pores. The presence of these pores was attributed to the adsorption of Pb(II) ions to the surface of EYMC. As a result, the adsorbent displayed maximum Pb(II) adsorption at pH 4.0–6.0 and with high adsorption capacity 121.6 mg/g.

Thiocarbohydrazide-modified chitosan (TCHECS) was synthesized for the removal of As(V), Ni(II), Cu(II), Cd(II), Pb(II) ions in aqueous systems [57]. The developed adsorbent showed the ability to form pH dependent gels. They observed 55.6–99.0% removal at pH 9.0. Further, they have developed two new chitosan-grafted materials as thiosemicarbazide (TSFCS) and thiocarbohydrazide (TCFCS) based on a similar concept [58]. And prepared adsorbents were also tested for the similar ions. The adsorption efficiency of TSFCS- and TCFCS-grafted chitosan was 66.4–99.9% and 71.5–99.9%, correspondingly, which is larger than TCHECS.

In another study, structural modification of chitosan was carried out by Monier to develop chitosan–thioglyceraldehyde Schiff's base cross-linked magnetic resin

(CSTG) [59]. The synthesized resin presented an average pore size of 795 nm and BET surface area of 70.5 m^2/g. The developed resin was tested for Hg(II), Cu(II) and Zn(II) ions and showed 98, 76, 52 mg/g removal capacities, correspondingly. Negm et al. prepared modified chitosan materials as chloroacetic-grafted chitosan and glycine-grafted chitosan [60]. The adsorbent was prepared by mixing of equivalent molar concentration of chitosan and glycine/chloroacetic acid in xylene solvent heated at 130 °C for 3 h. The reaction was forced stop when the equivalent quantity of water was attained. The chloroacetic-grafted chitosan showed adsorption capacity of 59.1 mg/g for Co(II), and 175.12 mg/g for Cu(II), whereas glycine-grafted chitosan showed 82.9, 165.91 mg/g for Co(II), and Cu(II) ions, correspondingly.

To improve selectivity of chitosan, cross-linking of chitosan using glutaraldehyde (chitosan-GLA) and epichlorohydrin (chitosan-ECH) agents was done by Rabelo et al. [61]. The adsorbents were found to be selective for Cu(II) and Hg(II) ions in aqueous solutions. The resultant chitosan-GLA showed a higher adsorption capacity towards Cu(II) than Hg(II), whereas chitosan-ECH showed a higher adsorption capacity towards Hg(II) as compared to Cu(II) metal ions.

Amino-terminated hyper branched dendritic polyamide amine first generation (1ACB) chitosan beads were synthesized by Gandhi and Meenakshi subsequent to the grafting onto the chitosan beads [62]. The transformation was achieved in two steps. Firstly, Michael methyl acrylate addition to surface amino groups and secondly, terminal ester groups amidation using ethylene diamine. The complex modification reactions resulted in the development of additional second generation chitosan (2ACB) and third generation (3ACB) beads Further protonation and Zr (IV) loading of 3ACB resulted in enhancing Cr(VI) adsorption. Out of the three types of generated beads, the 3ACB beads showed a high adsorption capacity of 224.2 mg/g.

The effect of different interfering ions on adsorption of Co(II) by EDTA-modified chitosan was studied by Repo et al. [63]. The decrease in adsorption rate was observed with EDTA as a result of competition between surface chelating agents and solution phase. The material demonstrated a specific surface area of 0.71 m^2/g, total pore volume of 1.8×10^{-3} cm^3/g and average pore size of 610 Å. The adsorbent presented specific adsorption of Co(II) with low adsorption capacity of 1.35 mmol/g. Xanthates have the properties of easy preparation with low-cost reagents and enhanced stability with the metal complexes. Xanthate carboxymethyl was utilized for grafting of chitosan by Song et al. to prepare xanthated carboxymethyl chitosan (XCC) for the selective uptake of Cu(II) and Ni(II) ions from aqueous solutions [64]. The introduction of both carboxyl and xanthate moieties into the bone of chitosan resulted in an enhanced adsorption capacity as compared to raw chitosan. The grafted chitosan adsorbent showed a higher adsorption capacity towards Cu(II) as compared to Ni(II) metal ions.

Kumar et al. carried out the microwave-assisted grafting of chitosan using n-butylacrylate onto for selective Cr(VI) uptake [65]. The presence of ester –C=O group peak at 1727 cm^{-1} in FTIR spectrum confirmed the formation of grafted derivative. The equilibrium isotherms studies were done using Dubinin–

Radushkevich, Temkin, Elovich and Redlich models. The adsorbent showed 17.15 mg/g adsorption capacity towards Cr(VI). The adsorption of Cr(VI) involved the interaction of the hydroxyl and amino moieties of chitosan and Cr(VI) ions.

Cross-linking of chitosan leads to its enhanced stability in acidic media, yet it reduces its adsorption efficiency due to the lesser availability of active functional groups because of cross-linking network. The problem was overcome by grafting new active functional groups on the cross-linked chitosan. Grafting of citric acid can result in the potential availability of its functional groups as active sites of heavy metals adsorption. Therefore, chitosan flakes were surface-modified using citric acid and cross-linking with glutaraldehyde [66]. The synthesized chitosan flakes carried out the selective removal of Pb(II) with an adsorption capacity of 101.7 mg/g. The adsorption experiments were carried out at pH 5 so as to evade the precipitation of $Pb(OH)_2$. Wang et al. modified chitosan with montmorillonite (CTS–MMT) for the selective removal of Co(II) [67]. The synthesized CTS–MMT adsorbent has 150 mg/g adsorption capacity towards Co(II). The grafting amine functional groups onto chitosan can demonstrate potential metal ions adsorption. Therefore, Xu et al. prepared triethylene–tetramine modified magnetic chitosan (TETA–MCS) for the selective removal of the Th(IV) from water [68]. The adsorbent presented removal capacity of 133.3 mg/g at normal room temperature. The developed magnetic adsorbent presented excellent removal capacity towards the Th(IV) because of numerous active sites and small reins size.

Utilization of chelating resins having appropriate functional groups for chelating chitosan may function as promising adsorbent for efficient metal ions adsorption. Further, the formation of chelate rings by polyamine can enhance the complex stability. Hence, the utilization of diethylenetriamine as polyamine can result in the coordination of U(VI) ions via a five-membered chelating rings, and therefore, high adsorption capacity. These properties were utilized by the same authors to prepare diethylenetriamine-functionalized magnetic chitosan for U(VI) adsorption with a maximum adsorption capacity of 65.16 mg/g [69]. Similarly, modification of magnetic chitosan using α-ketoglutaric acid (α-KA-Fe_3O_4/CS) was achieved by Yang et al. for removal of Cd(II) from aqueous solution [70]. The adsorbent presented maximum adsorption capacity of 201.2 mg/g at normal room temperature. Similarly, Kyzas and Deliyanni prepared magnetic chitosan for selective Hg(II) removal [71]. The adsorbent presented maximum adsorption capacity of 152 mg/g at pH 5 much higher than α-KA-Fe_3O_4/CS.

Further metal (lead or cadmium) uptake of chitosan was enhanced by similar authors by grafting it with poly(itaconic acid) (CS-g-IA) [72]. The authors prepared modified chitosan adsorbents by grafting with itaconic acid (CS-g-IA) and cross-linked either with glutaraldehyde [CS-g-IA(G)] or epichlorohydrin [CS-g-IA(E)] for the selective Cd(II) or Pb(II) uptake. The CS-g-IA(G) and CS-g-IA(E) demonstrated very high adsorption capacity of 405 and 331 mg/g for Cd(II) and Pb(II) ions, respectively. Ungrafted chitosan showed a lower adsorption capacity as compared to the grafted one [72]. The adsorbents showed better chemical and biological degradation resistance and efficient adsorption capacity under extreme pH conditions as compared to magnetic chitosan [71].

The pH dependency of the metal ions adsorption on chitosan was demonstrated by Verbych et al. For that the authors utilized chitosan flakes for the selective uptake of Cu(II) from water [73]. The chitosan flakes showed a capacity of 1.8–2.2 mmol/g dry mass for Cu(II) adsorption. A sharp increase in capacity was observed in high chloride ions concentration. The variation of solution pH leads to competition between the coordination of Cu(II) with chitosan. A sharp increase in adsorption capacity for copper was found at pH 5.4–6.0. It was due to the alteration in the number of $-NH_2$ groups for copper ion binding with solution pH. On increasing pH of the solution around pH 5–6, the number of $-NH_2$ groups available for copper ion binding were highest. Therefore, the optimal pH range for copper adsorption was from 5.4–6.0. Mcafee et al. utilized the commercial chitin, chitosan and chitosan cross-linked with benzoquinone for heavy metal ions detoxification [74]. Cross-linking of chitosan in polymeric matrices has been found to increase both mechanical and chemical stability. Therefore, benzoquinone was utilized to develop benzoquinone cross-linked chitosan. The highest adsorption capacity of 137, 108, 58 and 124 mg/g was achieved for copper, zinc, arsenic and chromium, correspondingly. Cross-linked chitosan beads have been chemically modified in various ways to increase the removal performance [75]. From them, the reaction of ethylenediamine and carbodiimide resulted in the formation of aminated chitosan beads that showed special affinity for Hg(II) ions. The developed beads showed uptake capacity of 2.26 mmol Hg(II)/g dry mass at pH 7.0 which is highest among different bioadsorbents. Miretzky and Cirelli extensively reviewed the chitosan and its derivatives based adsorbents for Hg(II) removal from water [76]. It was found that modified chitosan has various benefits as compared to powdered chitosan like superior surface area, and cross-linking in modified beads resulting in their insolubility at low pHs. The coating of chitosan onto perlite ore resulted in the development of novel chitosan biosorbent which was successfully utilized for Cu(II) and Ni(II) uptake [77]. The developed chitosan biosorbent showed the higher monolayer adsorption capacity for Cu(II) as compared to Ni(II) at pH 5.0. However, the adsorbent showed the lower adsorption capacity for both Cu(II) and Ni(II) as compared to XCC adsorbent. This is due to the presence of the carboxyl groups and xanthate groups in XCC that enhanced the adsorption capacity of the XCC adsorbent.

Paulino et al. utilized silkworm chrysalides (ChSC) derived chitosan for the selective uptake of Pb(II) and Cu(II) from industrial wastewater [78]. The adsorbent showed best pure ChSC deacetylation degrees (DDs) as 80% (90 min ChSC deacetylation) for Pb(II) uptake and 92% (180 min ChSC deacetylation) for Cu(II). The pure ChSC with 80% DD showed 72 mg/g capacities for Pb(II) and ChSC with 80% DD showed 87 mg/g capacity for Cu(II) at pH 5.0.

Krishnapriya and Kandaswamy developed a new chitosan derivative by cross-linking a metal complexing agent, [6,6′-piperazine-1,4-diyldimethylenebis (4-methyl-2-formyl) phenol] (L), with chitosan (CTS) [79]. The results showed adsorption capacities of for the various metal in the order Cu(II) > Ni(II) > Cd (II) $\geq$ Co(II) $\geq$ Mn(II) $\geq$ Fe(II) $\geq$ Pb(II) in the pH range 6.5–8.5. Polyvinyl chloride (PVC) is an abundantly available biopolymer in nature. Therefore, Popuri et al. developed biosorbent by coating chitosan onto polyvinyl chloride

(PVC) beads [80] for Cu(II) and Ni(II) uptake. The results demonstrated maximum uptake capacity of 87.9 mg/g for Cu(II) at pH 4.0 and 120.5 mg/g for Ni(II) ions at pH 5.0, correspondingly. The reported adsorption capacity of the present adsorbent was found to be lower than XCC [64] and chitosan-coated perlite adsorbent [77].

Liu et al. synthesized magnetic chitosan nanocomposites on account of amine-functionalized magnetite nanoparticles [81]. These nanocomposites presented rapid Pb(II), Cu(II) and Cd(II) uptake from water. The synthesized magnetic chitosan nanocomposites can be recycled for the repeated removal of heavy metal ions. For that the developed adsorbent has been easily recovered from water using an external magnetic field. The uptake capacity of 45.45 mg/g was achieved for Al(III) from aqueous solutions in the pH between 3.0–4.0 at 30 °C. [82].

Chitosan was utilized by Guzman et al. for the very efficient removal of vanadate anions from dilute solutions [83]. The adsorbent displayed sorption capacity of 400–450 mg/g in the pH range of 3.0–3.5. The prevalence of anionic groups specified that sorption took place by their action of protonated amine sites with decavanadate ions. The high proportion of non-adsorbable vanadate ions in near-neutral solutions ended with minor adsorption. A significant reduction in protonation of amine groups occurred above pH 7.0 that resulted in the inability of polymer to exchange counter anions with vanadate moieties. Regeneration of adsorbent was achieved in alkaline solutions, where vanadium was presented in the form of vanadate ions in solution.

A lot of research has been carried out to remove arsenic from water and wastewater systems. Molybdate-impregnated chitosan gel beads were also utilized by Dambies et al. [84] for the sorption of As(V) from drinking water. The molybdate impregnation on unmodified chitosan increased the adsorption capacity for As(V). The modified adsorbent showed 160 mg/g sorption capacity at pH 3.0. Shrimp shells derived chitosan powder, an agricultural waste was utilized in the form of beads for the adsorption of As(III) and As(V) from water [85]. The adsorbent presented low adsorption capacities as 1.83 and 1.94 mg As/g for As(III) and As(V), respectively, at pH 5. Similarly, Kwok et al. [86] utilized chitosan flakes for the adsorption of As(V). An increase in pH from 3.5 to 4.5 increased desorption rate of arsenate from chitosan. The adsorption capacity towards arsenate ion was a function of the protonation reaction of chitosan. A reduction in concentration of protonated groups on chitosan, available for arsenate sorption occurred with the increase in pH of the aqueous phase. The arsenic ions in the aqueous phase were presented as $H_2AsO_4^-$. The adsorbent showed higher adsorption capacity (14.160 mg arsenate/g chitosan) at initial pH 3.5 as compared to pH 5.5. Gupta et al. utilized iron–chitosan composites for the removal of As(III) and arsenic(V) from groundwater [87]. The adsorption studies demonstrated higher removal capacity by iron chitosan flakes (ICF) [22.47 mg/g for As(V) and 16.15 mg/g for As(III)] as compared to chitosan granules (ICB) [2.24 mg/g for As(V) and 2.32 mg/g for As(III)]. The presence of various anions like sulphate, phosphate and silicate caused no significant interference in arsenic uptake. Both ICF and ICB adsorbents were employed in column regeneration studies for two successive regeneration cycles during arsenic removal.

The wastewater generated in various industries like textile, electroplating, leather tanning and metallurgy industries are rich in Cr(VI) and Cr(III). Hexavalent form of chromium is more lethal as compared to the trivalent one. Therefore, it is imperative to remove chromium and bring down its concentration at permissible level. In this regard, Aydın and Aksoy [88] carried out adsorption of Cr(VI) onto chitosan flakes. The effect of various process parameters like pH (1.5–9.5), adsorbent dose (1.8–24.2 g/L) and initial concentration (15–55 mg/L) was studied. The adsorbent demonstrated 22.09 mg/g adsorption capacity sat initial chromium concentration of 30 mg/L with the adsorbent dosage of 13 g/L at pH 3.0 which increased to 102 mg/g for 100 mg/L initial Cr(VI) concentration. Hasan et al. enhanced the adsorption capacity of chitosan for Cr(VI) by coating it by an inert substrate, perlite. Chitosan-coated perlite beads were prepared by drop by drop addition of liquid chitosan slurry and perlite to an alkaline medium [89]. The adsorbent showed 104 mg/g adsorption capacity for the initial concentration of Cr(VI) as 5000 mg/L. The regeneration of spent beads was achieved by different concentrations of NaOH solution.

In a study, Cr(VI) adsorption from synthetic and actual wastewaters was achieved at 25 °C by a composite chitosan biosorbent which was prepared by coating chitosan onto non-porous ceramic alumina. [90]. The twice-coated biosorbent showed higher adsorption capacities as compared to other reported in literature due to the coating of the chitosan. Further, the increased surface area and the generation of high concentration of chromium binding sites on chitosan attributed the high adsorption capacity.

Spinelli et al. utilized quaternary ammonium salt of chitosan (QCS), prepared by reaction of a quaternary trimethylammonium, and glycidyl chloride for the removal of Cr(VI) [91]. The adsorption was most favourable at pH 3.5–4.5 because of the presence of Cr(VI) in the form of $HCrO_4^-$. The adsorption capacity of Cr(VI) reduced from 68.3 to 30.2 mg/g on increasing pH from 4.5 to 9.0. The elution of Cr(VI) ions from the spent QCS was achieved using 1 mol/L solution of NaCl/NaOH without any significant loss in adsorption capacity [92].

In another study, chitosan was utilized as a stabilizer for the development of chitosan-Fe(0) nanoparticles [chitosan-Fe(0)]. Effect of various experimental parameters like the initial Cr(VI) concentration, adsorbent dose on Cr(VI) adsorption by chitosan-Fe(0) was studied in batch experiments [93]. Cr(VI) removed by both physical adsorption of Cr(VI) onto the adsorbent surface and successive Cr(VI) reduction to Cr(III). High-resolution X-ray photoelectron spectroscopy studies demonstrated the predominant presence of Cr(III) and Fe(III) species as compared to Cr(VI) and Fe(0) on the surface of adsorbent after the reaction. Various modified chitosan-based adsorbents for heavy metal ions removal are listed in Table 1.

Yan et al. prepared aminated chitosan adsorbent for heavy metal ions adsorption [99]. The amount of amidocyanogenon adsorbent was increased four times as compared to chitosan cross-linked adsorbent by cross-linking amination reaction. The resultant cross-linked adsorbent showed enhanced adsorption ability towards nickel citrate and Cr(VI). The aminated chitosan adsorbent showed adsorption capacity of up to 30.2 mg/g for nickel citrate and 28.7 mg/g for Cr(VI) at the initial

Table 1 Various chitosan and modified chitosan-based adsorbents for heavy metal ions removal

Adsorbent	Metal ion	Removal capacity (mg/g)	References
Chitosan	Al(III)	45.45	[82]
Chitosan	VO_4^{3-}	400–450	[83]
Chitosan beads	As(III), As(V)	1.83, 1.94	[85]
Chitosan coated onto non-porous ceramic alumina	Cr(VI)	154.0	[90]
Cross-linked chitosan	Cr(VI)	215.0	[92]
Chitosan	Cd(II)	5.93	[94]
Chitosan	Hg(II), Cu(II), Ni(II), Zn(II)	815, 222, 164, 75	[95]
Porous-magnetic chitosan beads	Cd(II)	188–518	[96]
Carboxyl-grafted chitosan	Cu(II)	318.0	[97]
Glutaraldehyde cross-linked chitosan beads	MoO_4^-	763.0	[98]
Aminated chitosan	Ni(II), Cr(VI)	30.2, 28.7	[99]
Cross-linked chitosan	U(VI)	72.6	[100]
Chitosan	Ag	42.0	[101]
Chitosan	Au(III)	30.95	[102]
Chitosan benzoyl thiourea derivative	Co(II)	29.47	[103]

metal ion concentration of 1000 mg/L. Lasko et al. utilized chitosan for the uptake of silver from engineering wastewaters [101]. Batch as well as column studies demonstrated the adsorption of hydrated silver ion in addition to ammonia, thiocyanate, thiosulfate and cyanide complexes of silver from synthetic wastewater. The adsorption of free silver ion, Ag^+, more particularly $[Ag(NH_3)_2]^+$, occurred at broad pH range of 4.0–8.0. The adsorbent presented 42 mg/g adsorption capacity for silver ions in a continuous column treatment at pH 6.0. The inverse phase emulsion dispersion technique was utilized by Zhou et al. for the preparation of chitosan microparticles which were further modified with thiourea (TCS) [104]. The developed TCS adsorbent presented efficient and selective adsorption of Pt(IV) and Pd(II) at pH 2.0 in the presence of Cu(II), Pb(II), Cd(II), Zn(II), Ca(II) and Mg(II) ions. The applicability of pseudo-second-order kinetics demonstrated chemical adsorption as the main adsorption mechanism. The adsorbent showed the Langmuir adsorption capacity of 129.9 mg/g and 112.4 mg/g for Pt(IV) and for Pd(II), correspondingly. The regenerated adsorbent can be utilized up to 5 cycles without any significant loss in adsorption capacity and the adsorbed metal ions were successfully recovered with 97% efficiency. The above-discussed literature studies have proven that chitosan-based natural sorbents are very potential biosorbents for metal ions adsorption. Further, the adsorption of these heavy metal ions on chitosan-based adsorbents has been known to occur through ion exchange mechanism in acidic media, metal chelation and as a result of ion pairs formation [105, 106]. A number of factors are known to control this reaction, for example, adsorbent charge, pH of the solution and the hydrolyzing ability of metal ions and the ability of polynuclear species formation [105–109].

5 Summary

The present chapter aims at the current progress associated with the desalination and heavy metal ion detoxification of water and wastewater using chitosan-based natural biosorbents over the previous 10–15 years. The adsorption capacities demonstrated in the various publications present effectiveness of the sorbent for the specific type of metal species which in turn depends on various experimental parameters. The utilization of chitosan-based sorbents for water and wastewater desalination and heavy metal detoxification presents excellent adsorption capacity towards metal ions, economic viability, non-hazardous nature, and biocompatibility. However, the swelling of chitosan is known as a serious drawback which prevents its utilization at large scale. Still, there are some industries that utilize industrial grade chitosan for water and wastewater desalination and heavy metal detoxification. Therefore, the potential of modified materials is still to be explored for the large-scale utilization of these biosorbents.

6 Future Scenario

Although extensive studies in literature on the utilization of chitosan biosorbents for desalination and heavy metal ion detoxification of water and wastewater have been done, still there are a number of research gaps that require to be filled. Some of the vital characteristics that need to be addressed are summed up as:

(1) The prime matter is to choose a suitable variety of chitosan adsorbent to attain the highest adsorption of a particular contaminant according to the adsorbent–adsorbate interactions.
(2) To enhance the removal efficiency towards various pollutants, the optimization of various conditions leading to chitosan with high amino groups on its surface is required.
(3) Chitosan-based sorbents with low fabrication cost and high adsorption efficiency should be encouraged.
(4) Proper mechanistic details of organic and inorganic contaminants with chitin and chitosan derivatives are needed to offer an exact binding mechanism.
(5) For the enhanced fiscal viability of the method, detailed regeneration studies should be carried out with the used chitosan derivatives in order to recover the metals with adsorbent.

(6) Multicomponent contaminants treatment potential of chitosan derivatives is required for the large-scale utilization of chitosan.
(7) In addition to the effect of various co-contaminants, more experiments should be performed on the effect of the presence of phenols, dyes on adsorption of metal ions.
(8) In addition to lab-based batch studies, large-scale pilot plant studies are needed to test chitosan derivative's applications at industrial level.
(9) The process efficiency not only depends on the adsorbent and adsorbate characteristics but also on different experimental conditions. Therefore, these variables need to be considered for studying the efficiency of chitosan derivatives.

Therefore, development of the chitosan-based adsorbents having all the above-stated properties may present noteworthy benefits than presently developed commercially costly activated carbons.

Acknowledgements We gratefully acknowledge support from the Ministry of Human Resource Development Department of Higher Education, Government of India, under the scheme of Establishment of Centre of Excellence for Training and Research in Frontier Areas of Science and Technology (FAST), for providing the necessary financial support to carry out this study vide letter No, F. No. 5–5/201 4–TS.Vll.

References

1. Water, L.G.A., W.L. Driessen, M.W. Glenny, J. Reedijk, and M. Schroader. 2002. Selective and reversible extraction of heavy metal-ions by mixed-donor crown ether-modified oxirane and thiirane resins. *Reactive & Functional Polymers* 51: 33–47.
2. Ramos, V.M., N.M. Rodrıguez, M.S. Rodrıguez, A. Heras, and E. Agullo. 2003. Modified chitosan carrying phosphonic and alkyl groups. *Carbohydrate Polymers* 51: 425–429.
3. Kumar, M.N.V.R. 2000. A review of chitin and chitosan applications. *Reactive & Functional Polymers* 46: 1–27.
4. Sashiwa, H., and S. Aiba. 2004. Chemically modified chitin and chitosan as biomaterials. *Progress in Polymer Science* 29: 887–908.
5. Miyazaki, S., K. Ishii, and T. Nadai. 1981. The use of chitin and chitosan as drug carriers. *Chemical & Pharmaceutical Bulletin* 29: 3067–3069.
6. Chang, Y.-C., S.-W. Chang, and D.-H. Chen. 2006. Magnetic chitosan nanoparticles: studies on chitosan binding and adsorption of Co(II) ions. *Reactive & Functional Polymers* 66: 335–341.
7. Hsien, T.-Y., and G.L. Rorrer. 1995. Effects of acylation and crosslinking on the material properties and cadmium ion adsorption capacity of porous chitosan beads. *Separation Science and Technology* 30: 2455–2475.
8. Li, N., and R. Bai. 2005. Copper adsorption on chitosan-cellulose hydrogel beads: Behaviors and mechanisms. *Separation and Purification Technology* 42: 237–247.
9. Guibal, E. 2004. Interactions of metal ions with chitosan-based sorbents: A review. *Separation and Purification Technology* 38: 43–74.
10. Kurita, K., T. Sannan, and Y. Iwakura. 1979. Studies on chitin. VI. Binding of metal cations. *Journal of Applied Polymer Science* 23: 511–515.

11. Chaterjee, S., T. Chaterjee and S.H. Woo. 2010. A new type of chitosan hydrogel sorbent generated by anionic surfactant gelation. *Bioresource Technology* 101: 3853–3858.
12. Zhi, J., Y. Wang, and G. Luo. 2005. Adsorption of diuretic furosemide onto chitosan nanoparticles prepared with a water-in-oil nanoemulsion system. *Reactive & Functional Polymers* 65: 249–257.
13. Haider, S., and S.Y. Park. 2009. Preparation of the electrospun chitosan nanofibers and their applications to the adsorption of Cu(II) and Pb(II) ions from an aqueous solution. *Journal of Membrane Science* 328: 90–96.
14. Vieira, R., and M. Beppu. 2005. Mercury ion recovery using natural and crosslinked chitosan membranes. *Adsorption* 11: 731–736.
15. Davila–Rodriguez, J.L., V.A. Escobar–Barrios, K. Shirai, and J.R. Rangel–Mendez. 2009. Synthesis of a chitin-based biocomposite for water treatment: Optimization for fluoride removal. *Journal of fluorine Chemistry* 130: 718–726.
16. Kamble, S.P., S. Jagtap, N.K. Labhsetwar, D. Thakare, S. Godfrey, S. Devotta, and S.S. Rayalu. 2007. Defluoridation of drinking water using chitin, chitosan and lanthanum-modified chitosan. *Chemical Engineering Journal* 129: 173–180.
17. Sundaram, C.S., N. Viswanathan, and S. Meenakshi. 2008. Uptake of fluoride by nano-hydroxyapatite/chitosan, a bioinorganic composite. *Bioresource Technology* 99: 8226–8230.
18. Sundaram, C.S., N. Viswanathan, and S. Meenakshi. 2009. Fluoride sorption by nano-hydroxyapatite/chitin composite. *Journal of Hazardous Materials* 172: 147–151.
19. Sundaram, C.S., N. Viswanathan, and S. Meenakshi. 2009. Defluoridation of water using magnesia/chitosan composite. *Journal of Hazardous Materials* 163: 618–624.
20. Jagtap, S., D. Thakre, S. Wanjari, S. Kamble, N. Labhsetwar, and S. Rayalu. 2009. New modified chitosan-based adsorbent for defluoridation of water. *Journal of Colloid and Interface Science* 332: 280–290.
21. Swain, S.K., R.K. Dey, M. Islam, R.K. Patel, U. Jha, T. Patnaik, and C. Airoldi. 2009. Removal of fluoride from aqueous solution using aluminum-impregnated chitosan biopolymer. *Separation Science and Technology* 44: 2096–2116.
22. Thakre, D., S. Jagtap, A. Bansiwal, N. Labhsetwar, and S. Rayalu. 2010. Synthesis of La-incorporated chitosan beads for fluoride removal from water. *Journal of Fluorine Chemistry* 13: 373–377.
23. Thakre, D., S. Jagtap, A. Bansiwal, N. Labhsetwar, and S. Rayalu. 2010. Chitosan based mesoporous Ti–Al binary metal oxide supported beads for defluoridation of water. *Chemical Engineering Journal* 158: 315–324.
24. Viswanathan, N., and S. Meenakshi. 2008. Selective sorption of fluoride using Fe(III) loaded carboxylated chitosan beads. *Journal of Fluorine Chemistry* 129: 503–509.
25. Viswanathan, N., and S. Meenakshi. 2008. Enhanced fluoride sorption using La(III) incorporated carboxylated chitosan beads. *Journal of Colloid and Interface Science* 322: 375–383.
26. Viswanathan, N., and S. Meenakshi. 2009. Synthesis of Zr(IV) entrapped chitosan polymeric matrix for selective fluoride sorption. *Colloids and Surfaces* B72: 88–93.
27. Viswanathan, N., and S. Meenakshi. 2009. Enhanced and Selective Fluoride Sorption on Ce(III) Encapsulated Chitosan Polymeric Matrix. *Journal of Applied Polymer Science* 112: 1114–1121.
28. Viswanathan, N., and S. Meenakshi. 2010. Development of chitosan supported zirconium (IV) tungstophosphate composite for fluoride removal. *Journal of Hazardous Materials* 176: 459–465.
29. Viswanathan, N., and S. Meenakshi. 2010. Selective fluoride adsorption by a hydrotalcite/chitosan composite. *Applied Clay Science* 48: 607–611.
30. Viswanathan, N., and S. Meenakshi. 2010. Enriched fluoride sorption using alumina/chitosan composite. *Journal of Hazardous Materials* 178: 226–232.

31. Sorlier, P., A. Denuziere, C. Viton, and A. Domard. 2001. Relation between the degree of acetylation and the electrostatic properties of chitin and chitosan. *Biomacromolecules* 2: 765–772.
32. Yui, T., K. Imada, K. Okuyama, Y. Obata, K. Suzuki, and K. Ogawa. 1994. Molecular and crystal-structure of the anhydrous form of chitosan. *Macromolecules* 27: 7601–7605.
33. Rinaudo, M. 2006. Characterization and properties of some polysaccharides used as biomaterials. *Macromolecular Symposia* 245–246: 549–557.
34. Crini, G. 2005. Recent developments in polysaccharide-based materials used as adsorbents in wastewater treatment. *Progress in Polymer Science* 30: 38–70.
35. Kurita, K. 2006. Chitin and chitosan: Functional biopolymers from marine crustaceans. *Marine Biotechnology* 8: 203–226.
36. Hsien, T.-Y., and G.L. Rorrer. 1995. Effects of acylation and crosslinking on the material properties and cadmium ion adsorption capacity of porous chitosan beads. *Separation Science and Technology* 30: 2455–2475.
37. Huang, R., G. Chen, M. Sun, and C. Gao. 2008. Preparation and characterization of quaterinized chitosan/poly(acrylonitrile) composite nanofiltration membrane from anhydride mixture cross-linking. *Separation and Purification Technology* 58: 393–399.
38. Juang, R.S., and R.C. Shiau. 2000. Metal removal from aqueous solutions using chitosan-enhanced membrane filtration. *Journal of Membrane Science* 165: 159–167.
39. EL-Gendi, A., A. Deratani, S. A. Ahmed, and S. S. Ali. 2014. Development of polyamide-6/chitosan membranes for desalination. *Egyptian Journal of Petroleum* 23: 169–173.
40. Raval, H.D., P.S. Rana, and S. Maiti. 2015. A novel high-flux, thin-film composite reverse osmosis membrane modified by chitosan for advanced water treatment. *RSC Advances* 5: 6687–6694.
41. Li, J., Y. Xu, M. Hu, J. Shen, C. Gao, and B.V.D. Bruggen. 2015. Enhanced conductivity of monovalent cation exchange membranes with chitosan/pani composite modification. *RSC Advances* 5: 90969–90975.
42. Padaki, M., A.M. Isloor, J. Fernandes, and K.N. Prabhu. 2011. New polypropylene supported chitosan NF-membrane for desalination application. *Desalination* 280: 419–423.
43. Kumar, R., A.M. Isloor, A.F. Ismail, S.A. Rashid, and T. Matsuura. 2013. Polysulfone–chitosan blend ultrafiltration membranes: Preparation, characterization, permeation and antifouling properties. *RSC Advances* 3: 7855–7861.
44. Ayoub, A., R.A. Venditti, J.J. Pawlak, A. Salam, and M.A. Hubbe. 2013. Novel hemicellulose-chitosan biosorbent for water desalination and heavy metal removal. *ACS Sustainable Chemistry & Engineering* 1: 1102–1109.
45. Chen, Y., L. Chen, H. Bai, and L. Li. 2013. Graphene oxide–chitosan composite hydrogels as broad-spectrum adsorbents for water purification. *Journal of Materials Chemistry A* 1: 1992–2001.
46. Sorour, M.H., H.A. Hani, H.F. Shaalan, M.M. El-Sayed, and M.M.H. El-Sayed. 2015. Softening of seawater and desalination brines using grafted polysaccharide hydrogels. *Desalination and Water Treatment* 55: 2389–2397.
47. Arvand, M., and M.A. Pakseresht. 2013. Cadmium adsorption on modified chitosan-coated bentonite: Batch experimental studies. *Journal of Chemical Technology and Biotechnology* 88: 572–578.
48. Chen, J.H., D.Q. Lu, B. Chen, and P.K. Ouyang. 2013. Removal of U(VI) from aqueous solutions by using MWCNTs and chitosan modified MWCNTs. *Journal of Radioanalytical and Nuclear Chemistry* 295: 2233–2241.
49. Chen, Y., and J. Wang. 2012. The characteristics and mechanism of Co(II) removal from aqueous solution by a novel xanthate-modified magnetic chitosan. *Nuclear Engineering and Design* 242: 452–457.
50. Debnath, S., A. Maity, and K. Pillay. 2014. Magnetic chitosan–GO nanocomposite: Synthesis, characterization and batch adsorber design for Cr(VI) removal. *Journal of Environmental Chemical Engineering* 2: 963–973.

51. Hummers, W.S., and R.E. Offeman. 1958. Preparation of graphitic oxide. *Journal of the American Chemical Society* 80: 1339.
52. Elwakeel, K.Z., A.A. Atia, and E. Guibal. 2014. Fast removal of uranium from aqueous solutions using tetraethylenepentamine modified magnetic chitosan resin. *Bioresource Technology* 160: 107–114.
53. Eser, A., V. NüketTirtom, T. Aydemir, S. Becerik, and A. Dinçer. 2012. Removal of nickel (II) ions by histidine modified chitosan beads. *Chemical Engineering Journal* 210: 590–596.
54. Gandhi, M.R., G.N. Kousalya, and S. Meenakshi. 2012. Selective sorption of Fe(III) using modified forms of chitosan beads. *Journal of Applied Polymer Science* 24: 1858–1865.
55. Kandile, N.G., and A.S. Nasr. 2014. New hydrogels based on modified chitosan as metal biosorbent agents. *International Journal of Biological Macromolecules* 64: 328–333.
56. Li, T.T., Y.G. Liu, Q.Q. Peng, X.J. Hu, T. Liao, H. Wang, and M. Lu. 2013. Removal of lead(II) from aqueous solution with ethylenediamine-modified yeast biomass coated with magnetic chitosan microparticles: Kinetic and equilibrium modeling. *Chemical Engineering Journal* 214: 189–197.
57. Li, M.L., R.H. Li, J. Xu, X. Han, T.Y. Yao, and J. Wang. 2014. Thiocarbohydrazide-modified chitosan as anticorrosion and metal ion adsorbent. *Journal of Applied Polymer Science* 131: 8437–8443.
58. Li, M., J. Xu, R. Li, D. Wang, T. Li, M. Yuan, and J. Wang. 2014. Simple preparation of aminothiourea-modified chitosan as corrosion inhibitor and heavy metal ion adsorbent. *Journal of Colloid and Interface Science* 417: 131–136.
59. Monier, M. 2012. Adsorption of Hg2+, Cu2+ and Zn2+ ions from aqueous solution using formaldehyde cross-linked modified chitosan-thioglyceraldehyde schiff's base. *International Journal of Biological Macromolecules* 50: 773–781.
60. Negm, N.A., R. Sheikh, A.F. El-Farargy, H.H.H. Hefni, and M. Bekhit. 2015. Treatment of industrial wastewater containing copper and cobalt ions using modified chitosan. *Journal of Industrial and Engineering Chemistry* 21: 526–534.
61. Rabelo, R., R. Vieira, F. Luna, E. Guibal, and M. Beppu 2012. Adsorption of copper (II) and mercury (II) ions on to chemically-modified chitosan membranes: equilibrium and kinetic properties. *Adsorption Science & Technology* 30: 1–21.
62. Gandhi, M.R., and S. Meenakshi. 2013. Preparation of amino terminated polyamidoamine functionalized chitosan beads and its Cr(VI) uptake studies. *Carbohydrate Polymers* 91: 631–637.
63. Repo, E., R. Koivula, R. Harjula, and M. Sillanpää. 2013. Effect of EDTA and some other interfering species on the adsorption of Co(II) by EDTA-modified chitosan. *Desalination* 321: 93–102.
64. Song, Q., C. Wang, Z. Zhang, and J. Gao. 2014. Adsorption of Cu(II) and Ni(II) using a novel xanthated carboxymethyl chitosan. *Separation Science and Technology* 49: 1235–1243.
65. Kumar, A.S.K., C.U. Kumar, V. Rajesh, and N. Rajesh. 2014. Microwave assisted preparation of N-Butylacrylate grafted chitosan and its application for Cr(VI) adsorption. *International Journal of Biological Macromolecules* 66: 135–143.
66. Suc, N.V., and H.T.Y. Ly. 2013. Lead (II) removal from aqueous solution by chitosan flake modified with citric acid via crosslinking with glutaraldehyde. *Journal of Chemical Technology and Biotechnology* 88: 1641–1649.
67. Wang, H., H. Tang, Z. Liu, X. Zhang, Z. Hao, and Z. Liu. 2014. Removal of cobalt(II) ion from aqueous solution by chitosan-montmorillonite. *Journal of Environmental Sciences* 26: 1879–1884.
68. Xu, J., L. Zhou, Y. Jia, Z. Liu, and A.A. Adesina. 2014. Adsorption of thorium (IV) ions from aqueous solution by magnetic chitosan resins modified with triethylene-tetramine. *Journal of Radioanalytical and Nuclear Chemistry* 303: 347–356.
69. Xu, J., M. Chen, C. Zhang, and Z. Yi. 2014. Adsorption of uranium(VI) from aqueous solution by diethylenetriamine-functionalized magnetic chitosan. *Journal of Radioanalytical and Nuclear Chemistry* 2013 (298): 1375–1383.

70. Yang, G., L. Tang, X. Lei, G. Zeng, Y. Cai, X. Wei, Y. Zhou, S. Li, Y. Fang, and Y. Zhang. 2014. Cd(II) removal from aqueous solution by adsorption on α-ketoglutaric acid-modified magnetic chitosan. *Applied Surface Science* 292: 710–716.
71. Kyzas, G.Z., and E.A. Deliyanni. 2013. Mercury (II) removal with modified magnetic chitosan adsorbents. *Molecules* 18: 6193–6214.
72. Kyzas, G.Z., P.I. Siafaka, D.A. Lambropoulou, N.K. Lazaridis, and D.N. Bikiaris. 2014. Poly (Itaconic Acid)-grafted chitosan adsorbents with different cross-linking for Pb (II) and Cd (II) uptake. *Langmuir* 40: 120–131.
73. Verbych, S., M. Bryk, G. Chornokur, and B. Fuhr. 2005. Removal of copper(II) from aqueous solutions by chitosan adsorption. *Separation Science and Technology* 40: 1749–1759.
74. Mcafee, B.J., W.D. Gould, J.C. Nadeau, and A.C.A.D. Costa. 2001. Biosorption of metal ions using chitosan, chitin, and biomass of rhizopus oryzae. *Separation Science and Technology* 36: 3207–3222.
75. Jeon, C., and W.H. Höll. 2003. Chemical modification of chitosan and equilibrium study for mercury ion removal. *Water Research* 37: 4770–4780.
76. Miretzky, P., and A.F. Cirelli. 2009. Hg(II) removal from water by chitosan and chitosan derivatives: A review. *Journal of Hazardous Materials* 167: 10–23.
77. Kalyani, S., J.A. Priya, P.S. Rao, and A. Krishnaiah. 2005. Removal of copper and nickel from aqueous solutions using chitosan coated on perlite as biosorbent. *Separation Science and Technology* 40: 1483–1495.
78. Paulino, A.T., L.B. Santos, and J. Nozaki. 2008. Removal of Pb^{2+}, Cu^{2+}, and Fe^{3+} from battery manufacture wastewater by chitosan produced from silkworm chrysalides as a low-cost adsorbent. *Reactive & Functional Polymers* 68: 634–642.
79. Krishnapriya, K.R., and M. Kandaswamy. 2009. Synthesis and characterization of a crosslinked chitosan derivative with a complexing agent and its adsorption studies toward metal(II) ions. *Carbohydrate Research* 344: 1632–1638.
80. Popuri, S.R., Y. Vijaya, V.M. Boddu, and K. Abburi. 2009. Adsorptive removal of copper and nickel ions from water using chitosan coated PVC beads. *Bioresource Technology* 100: 194–199.
81. Liu, X., Q. Hu, Z. Fang, X. Zhang, and B. Zhang. 2009. Magnetic chitosan nanocomposites: A useful recyclable tool for heavy metal ion removal. *Langmuir* 25: 3–8.
82. Septhum, C., S. Rattanaphani, J.B. Bremner, and V. Rattanaphani. 2007. An adsorption study of Al(III) ions onto chitosan. *Journal of Hazardous Materials* 148: 185–191.
83. Guzman, J., I. Saucedo, R. Navarro, J. Revilla, and E. Guibal. 2002. Vanadium interactions with chitosan: influence of polymer protonation and metal speciation. *Langmuir* 18: 1567–1573.
84. Dambies, L., E. Guibal, and A. Roze. 2000. Arsenic(V) sorption on molybdate-impregnated chitosan beads. *Colloid Surface A* 170: 19–31.
85. Chen, C.-C., and Y.-C. Chung. 2006. Arsenic removal using a biopolymer chitosan sorbent. *Journal of Environmental Science and Health Part A Toxic/Hazardous Substances and Environmental Engineering* 41: 645–658.
86. Kwok, K.C.M., V.K.C. Lee, C. Gerente, and G. McKay. 2009. Novel model development for sorption of arsenate on chitosan. *Chemical Engineering Journal* 151: 122–133.
87. Gupta, A., V.S. Chauhan, and N. Sankararamakrishnan. 2009. Preparation and evaluation of iron-chitosan composites for removal of As(III) and As(V) from arsenic contaminated real life groundwater. *Water Research* 43: 3862–3870.
88. Aydın, Y.A., and N.D. Aksoy. 2009. Adsorption of chromium on chitosan: Optimization, kinetics and thermodynamics. *Chemical Engineering Journal* 51: 188–194.
89. Hasan, S., A. Krishnaiah, T.K. Ghosh, D.S. Viswanath, V.M. Boddu, and E.D. Smith. 2003. Adsorption of Chromium(VI) on Chitosan-Coated Perlite. *Separation Science and Technology* 38: 3775–3793.

90. Boddu, V.M., K. Abburi, J.L. Talbott, and E.D. Smith. 2003. Removal of hexavalent chromium from wastewater using a new composite chitosan biosorbent. *Environmental Science and Technology* 37: 4449–4456.
91. Spinelli, V.A., M.C.M. Laranjeira, and V.T. Fávere. 2004. Preparation and characterization of quaternary chitosan salt: Adsorption equilibrium of chromium(VI) ion. *Reactive & Functional Polymers* 61: 347–352.
92. Rojas, G., J. Silva, J.A. Flores, A. Rodriguez, M. Ly, and M. Maldonado. 2005. Adsorption of chromium onto cross-linked chitosan. *Separation and Purification Technology* 44: 31–36.
93. Geng, B., Z. Jin, T. Li, and X. Qi. 2009. Kinetics of hexavalent chromium removal from water by chitosan-Fe^0 nanoparticles. *Chemosphere* 75: 825–830.
94. Jha, I.N., L. Iyengar, and A.V.S.P. Rao. 1988. Removal of cadmium using chitosan. *Journal of Environmental Engineering* 114: 964–974.
95. McKay, G., H.S. Blair, and A. Findon. 1989. Equilibrium studies for the sorption of metal-ions onto chitosan. *Indian Journal of Chemistry A* 28: 356–360.
96. Rorrer, G.L., T.-Y. Hsien, and J.D. Way. 1993. Synthesis of porous-magnetic chitosan beads for removal of cadmium ions from waste water. *Industrial and Engineering Chemistry Research* 32: 2170–2178.
97. Schmuhl, R., H.M. Krieg, K. Keizer. 2001. Adsorption of Cu (II) and Cr (VI) ions by chitosan: Kinetics and equilibrium studies. *Water SA* 27: 1–8.
98. Guibal, E., C. Milot, and J.M. Tobin. 1998. Metal-anion sorption by chitosan beads: equilibrium and kinetic studies. *Industrial and Engineering Chemistry Research* 37: 1454–1463.
99. Yan, Z., S. Haijia, and T. Tianwei. 2007. Adsorption Behaviors of the aminated chitosan adsorbent. *Korean Journal of Chemical Engineering* 24: 1047–1052.
100. Wang, G., J. Liu, X. Wang, Z. Xie, and N. Deng. 2009. Adsorption of uranium (VI) from aqueous solution onto cross-linked chitosan. *Journal of Hazardous Materials* 168: 1053–1058.
101. Lasko, C.L., and M.P. Hurst. 1999. An investigation into the use of chitosan for the removal of soluble silver from industrial wastewater. *Environmental Science and Technology* 33: 3622–3626.
102. Ngah, W.S.W., and K.H. Liang. 1999. Adsorption of gold (III) ions onto chitosan and N-Carboxymethyl chitosan: Equilibrium studies. *Industrial and Engineering Chemistry Research* 38: 1411–1414.
103. Metwally, E., S.S. Elkholy, H.A.M. Salem, and M.Z. Elsabee. 2009. Sorption behavior of ^{60}Co and 152+ 154 Eu radionuclides onto chitosan derivatives. *Carbohydrate Polymers* 76: 622–631.
104. Zhou, L., J. Liu, and Z. Liu. 2009. Adsorption of platinum(IV) and palladium(II) from aqueous solution by thiourea-modified chitosan microspheres. *Journal of Hazardous Materials* 172: 439–446.
105. Gerente, C., V.K.C. Lee, P.L. Cloirec, and G. McKay. 2007. Application of chitosan for the removal of metals from wastewaters by adsorption—Mechanisms and models review. *Critical Reviews in Environment Science and Technology* 37: 41–127.
106. Crini, G., and P.-M. Badot. 2008. Application of chitosan, a natural aminopolysaccharide, for dye removal from aqueous solutions by adsorption processes using batch studies: A review of recent literature. *Progress in Polymer Science* 33: 399–447.
107. Guibal, E. 2004. Interactions of metal ions with chitosan-based sorbents: A review. *Separation and Purification Technology* 38: 43–74.
108. Varma, A.J., S.V. Deshpande, and J.F. Kennedy. 2004. Metal complexation by chitosan and its derivatives: A review. *Carbohydrate Polymers* 55: 77–93.
109. Kyzas, G.Z., M. Kostoglou and N.K. Lazaridis. 2009. Copper and chromium(VI) removal by chitosan derivatives—Equilibrium and kinetic studies. *Chemical Engineering Journal* 152: 440–448.

Application of Biomaterials for Elimination of Damaging Contaminants from Aqueous Media

Vaishali Tomar and Dinesh Kumar

Abstract Natural materials are plentifully accessible low cost. A natural resource which is nontoxic to the ecosystem. Because of the excess amount of inorganic pollutants, organic pollutants and pathogens in water, it is harmful to human being. These contaminants should be taken out by the natural adsorbent due to the harmful force of these contaminants. This chapter surveys the current evolution of natural clays and their modified forms as adsorbing agents for treating drinking water. This chapter explores the adaptable nature of natural materials and nanomaterials with their capability to absorb multiplicity of contaminants, which are present in the drinking water. The properties and alteration of the natural adsorbent and its significance in removing a detailed type of contaminants are identified. The efficacy of the natural and modified adsorbents is compared to active technologies, materials and methods, and it is considerably higher or similar.

Keywords Water · Contaminants · Adsorption · Removal · Clean water

1 Introduction

Fresh drinking water is unrivaled of the important requisites for a healthy human person. Referable to the increasing industrialization, the function of chemicals has been increased the burden of unnecessary pollutants of drinking water. After so much research, it was found that heap of toxic materials is coming into soils and water. The accretion of heavy metals and metalloids is increasing in industrial countries, mine tailings, and it is harmful to health. For the removal of high metal wastes, there are tons of materials have been studied which we will talk about in

---

V. Tomar
Formulation, ABH Natures Product. New York 11717, USA
e-mail: vaishali.aura08@gmail.com

D. Kumar (✉)
School of Chemical Sciences, Central University of Gujarat, Gandhinagar 382030, India
e-mail: dinesh.kumar@cug.ac.in

© Springer International Publishing AG 2018
S. Bhardwaj Mishra and A.K. Mishra (eds.), *Bio- and Nanosorbents from Natural Resources*, Springer Series on Polymer and Composite Materials,
https://doi.org/10.1007/978-3-319-68708-7_7

this chapter [1, 2]. There are different kinds of toxic materials found in ground water which should be removed from the water. These materials are categorized by topic areas:

1. Heavy metals
   Heavy metals like lead (Pb), chromium (Cr), arsenic (As), zinc (Zn), cadmium (Cd), copper (Cu), mercury (Hg), and nickel (Ni) constitute an ill-defined group of inorganic chemical hazards.
2. Organic contaminants.
3. Pathogens.

The entry of these harmful substances into the ecosystem is increasing day by day. It was considered that the presence of extra fluoride, arsenic, and natural organic matters, heavy metals, and a mixture of pathogens are the major reasons of various waterborne diseases. The sole path to preserve the safety of water bodies is to develop effective materials and purifying technologies to remove these chemicals from drinking water sources. One such beneficial and successful process for the purification of water is the use of natural and modified adsorbents. Dissimilar characters of natural adsorbents have been applied for treatment of contaminated drinking water. The muds and their modified composites have been found to be effective and better, in comparison of another low-cost adsorbent [3–5].

Granting to the World Health Organization norms, the permissible concentration of fluoride ions in drinking water should be below 1.5 ppm. Fluoride is useful, as it receives a favorable effect on teeth below this boundary. The extra amount of fluoride causes fluorosis of the teeth and bones. There is an urgent need of remediation of water using low-cost, biodegradable biopolymers like chitosan. At the international and national level in this field, a lot of attempts have been made to consolidate the studies [6].

Organic pollutants in the ecosystem have been the most important environmental problems in the world. By the study of literature, it was revealed that there has been a high increase in production and utilization of organic pollutants in the last few years so it is resulting in a big danger of pollution. It was studied that efficient techniques have been used for the removal of highly toxic organic compounds from water.

Adsorption is a very much effective and low-cost technique for the removal of inorganic, organic pollutants, and pathogens from water, and it produces very good treated sewage. This chapter highlighted the removal of inorganic and organic pollutants by using an adsorption method with different kinds of natural and artificial adsorbents [7].

Many researchers have given considerable attention aimed to the removal efficiency of organic pollutants by adsorption technique. It was studied that natural materials like clays and modified clay, due to their high surface area and molecular sieve structure, are very much effective adsorbents for the removal of contaminants. The need to remove pathogens from potable water supplies is long recognized.

Pathogenic contamination of water causes illness and contributing increasing rate of disease around the world, and giving most serious impact in the developing world.

In this chapter, we delineate the description of pathogen groups and present an overview of the approaches which was used to remove inorganic, organic, and pathogens from water. We focus on the effectiveness of improved modern technologies for harmful contaminant removal [8–10].

In the past few years, nanocomposites have increased attention, including studies on developing the composites as sorbents for non-ionic and anionic pollutants [11], organic pollutants [12], anionic herbicide [13], and atrazine [14]. Chitosan–montmorillonite composites have been well known [15–19]. This chapter reviews the current use of natural clay and its composites as an environmentally efficient adsorbent for removal of organic, inorganic, and pathogenic contaminants from drinking water and its sources.

The chapter section is divided into following four headings based on the type of contaminant removed from the clay and its composites: (1) heavy metal contaminants, (2) organic contaminant, and (3) pathogens as shown in Fig. 1.

Fig. 1 Removal of contaminants from water by adsorption

2 Category of Contaminants Removed by the Materials

2.1 Heavy Metals

The serious effects on the health of human beings, animals, and plants have been done by heavy metal contamination in drinking water resources. Currently, many researchers are working in this field for removing various metals present in the water by an appropriate solution. After research and experiment, it was observed that heavy metal pollution has turned out to be one of the most serious environmental troubles today and the action of heavy metals is of unique worry due to their insubordination and perseverance in the environment. In the recent years, a variety of methods for the removal of heavy metal from wastewater have been widely studied. For the removal of various metals like arsenic, iron, manganese, lead, cadmium, uranium, chromium, selenium, tungsten, and zinc from the environment, many natural bio-adsorbents have been widely used.

If we would discuss natural bio-adsorbent, then clays and their modified forms attract the discussion partly because of their easy availability and relatively less cost. It has been receiving broad attention recently for use as adsorbents of metal ions from aqueous medium [20]. It was observed that for the removal of heavy metals by natural clays and their modified forms like kaolinite and montmorillonite [21]. It was reported and studied that the alteration of the mentioned clays with different polyoxy cations of Zr^{4+}, Al^{3+}, Si^{4+}, Ti^{4+}, Fe^{3+}, Cr^{3+}, Ga^{3+}, and so forth. The adsorptions of toxic metals, namely, As, Cd, Cr, Co, Cu, Fe, Pb, Mn, Ni, Zn, and so forth, have been premeditated primarily. It was found that a modified form of clay montmorillonite has much higher metal adsorption capacity compared to that modified form of kaolinite. Both are a good example for the removal of metals like Cd [22], CO [2, 23], Fe [24], Pb [25], Mn, and Zn [26].

Adsorption of heavy metals from aqueous solutions, sawdust coated by polyaniline (SD/PAn), and polyaniline composites has been also other good examples. Some combination of clay with iron oxide composite was studied for adsorption of metal ions Ni^{2+}, Cu^{2+}, Cd^{2+}, and Zn^{2+} from aqueous solution [27]. If we talk about the metal adsorption capacity of bentonite clay, then it was found that in the presence of iron oxide, the adsorption capacity of the bentonite was increased. It was shown by a simple magnetic separation method; the adsorbents have the advantage to be easily removed from the medium after saturation is reached. By batch experiments, it was observed that the removal of lead and cadmium ions from aqueous solutions was carried out by beidellite which is a low-cost and environmentally friendly material [28]. A natural montmorillonite modified clay with biopolymer chitosan was investigated for the removal of tungsten from drinking water in Nevada, USA, [29, 30].

Another natural adsorbent like activated carbon, kaolin, bentonite, blast furnace slag, and fly ash was used for the removal of the lead and zinc ions from aqueous water [31, 32]. The removal of lead and zinc by adsorption was investigated with various effects of contact time, pH, and adsorbent dosage on the adsorbent. It was found that metakaolin and mostly montmorillonite pre-treated with Fe^{2+}, Fe^{3+}, Al^{3+}, and Mn^{2+} salts were used for the removal of As from the groundwater [34, 35]. Chitosan–montmorillonite composites were considered for the removal of selenium from water [36]. It was researched that Ti-pillared montmorillonite (Ti-MMT) was used for the adsorption of arsenic from aqueous solutions with the role of contact time, pH, temperature, coexisting ions, and ionic potency [37, 38].

Fluoride is also coming in toxic material and because of the heavy amount of water it generates fluorosis in human. Fluorosis has affected millions of people [39–41] and widespread in at least 25 countries across the globe. Fluoride is beneficial as well as harmful. When it presents within the permissible limit of 1.0–1.5 mgL^{-1}, then it is useful for calcification of dental enamels [42]. The high concentration of fluoride above 1.5 mgL^{-1} in drinking water [43] caused fluorosis.

A lot of low-cost materials like bentonite clay modified using magnesium chloride were used for the removal of fluoride from water [44, 45] and characterized by using XRD and SEM techniques. Defluoridation from aqueous solutions by zirconium loaded bentonite (ZLB) was studied [46] at pH 6, maximum adsorption of fluoride from aqueous solutions was found. Zeolites with modified clinoptilolite and chabazite has been used for the removal of metals like Pb(II), Cd(II), Zn(II), and Cu(II). Sphagnum peat moss was used for the removal of heavy metals from ground water. Chitin, which is a long-chain polymer of an N-acetylglucosamine, an imitative of glucose was studied for the removal of heavy metal from aqueous solution.

Nowadays, nanomaterials are also used for the removal of heavy metals from aqueous water. A very less amount is generally required for the removal of fluoride from water. The montmorillonite-supported magnetite nanoparticles were used as an adsorbent for the removal of heavy metal from aqueous solution. It was characterized by different techniques like X-ray diffraction, nitrogen adsorption, elemental analysis, differential scanning calorimetry, transmission electron microscopy, and X-ray photoelectron spectroscopy [47–49].

A novel crystalline and hybrid Fe–Ce–Ni nanoporous adsorbent was developed for fluoride removal. A novel and efficient analytical method for the removal of fluoride using Zr–Mn composite material has been developed for water samples. The adsorption was confirmed by the use of various techniques like X-ray diffraction (XRD), Brunauer, Emmett, and Teller (BET) and FTIR [50–62].

In many locations in the world, the groundwater pollution by nitrates has been a widespread problem and it was removed by adsorption method. It was researched that natural calcium bentonite [63] by acid thermo-activation using HCl and H_2SO_4 has been examined for the removal of nitrate from an aqueous solution. The application of natural material and their modified forms for the removal of heavy materials from the water is shown in Table 1.

Table 1 Appliance of natural materials and their modified forms in removing heavy contaminants from drinking water

Contaminant	Type of clay and modification	Efficiency	Refs.
Cadmium Chromium Cobalt Copper Iron Lead	Kaolinite and montmorillonite and their modified forms bentonite	–	[22] [2, 23] [24] [25]
Zinc nickel Copper Cadmium Zinc	Bentonite clay iron oxide composite	–	[26, 27]
Lead Cadmium	Beidellite	83.3–86.9 mg g^{-1} 42–45.6 mg g^{-1}	[28]
Tungsten	Montmorillonite coated with chitosan	23.9 mg g^{-1}	[29]
Uranium	Thermally activated bentonite (TAB)	196 mL g^{-1}	[30]
Lead and zinc	Bentonite	5 g L^{-1} and 20 g L^{-1}	[31]
Hexavalent chromium	Montmorillonite-supported magnetic nanoparticle	15.3 mg g^{-1}	[32]
Cobalt	Kaolinite and montmorillonite		[33]
Arsenic	Calcined kaolin and bentonite pre-treated with Fe(II), Fe(III), Al(III) and Mn(II)	92–99%	[34]
Cadmium Chromium Copper Mercury Lead Zinc	Mixed clay (illite, kaolinite, mixed layer minerals and nonclay mineral carbonate fluorapatite	85% 90% 50% 60% 100% 92%	[35]
Selenium	Chitosan–montmorillonite	18.4 mg g^{-1}	[36]
Arsenate and arsenite	Ti-pillared montmorillonite	Greater than 60%	[37]
Lead Nickel Cadmium Copper	Bentonite-methylene bis-acrylamide	1666.67 mg g^{-1} 270.27 mg g^{-1} 416.67 mg g^{-1} 222.2 mg g^{-1}	[38]
Fluoride	Magnesium-incorporated bentonite magnesium–bentonite manganese–bentonite calciums	95.45%	[45, 46]
Nitrates	Montmorillonite activated by hydrochloric acid	22.28%	[63]

2.2 Organic Contaminants

A lot of organic compounds called organic pollution, and it comes from domestic sewage, town overflow, industrial effluents, and agricultural wastewater. It was studied that organic contaminants have become one of the environmental problems,

Table 2 Appliance of natural materials and its composites in removing organic contaminants from drinking water

Contaminants	Type of clay and modification	Efficiency	Refs.
Dichloroacetic acid	Bentonite-based absorptive ozonation followed by catalytic oxidation by Fe^{3+}	92%	[74]
Carbon tetrachloride	Quaternary ammonium salt-modified bentonite	70%	[75]
Emerging contaminants: naproxen, salicylic acid, clofibric acid and carbamazepine	Inorganic–organic intercalated (IO) bentonites	2.69 μ mol g^{-1} 5.55 μ mol g^{-1}	[76]
Phenol	Bentonite modified with cationic surfactant, acetyl trimethyl ammonium bromide (CTAB)	333 mg g^{-1}	[77]
Humid acid and O-dichlorobenzene	Combined ozonation and bentonite coagulation	95% of HA and 74% of DCB	[78]
Algae removal	Montmorillonite KSF	100%	[79]
Blue-green algae (cyanobacterial microcystis aeruginosa)	Montmorillonite-Cu^{2+}/Fe^{3+} oxides magnetic material	92%	[80]
Atrazine	4-vinylpyridine-co-styrene montmorillonite	90–99%	[81]
Atrazine, sulfentrazone, imazaquin, and alachlor	Vesicle–clay complex (Didodecyldimethylammonium bromide montmorillonite)	60% atrazine and 90–100% for others	[14]
Naphthalene and phenolic derivative	Crystal violet tetraphenyl phosphonium montmorillonite	99%	[82]
Salicylic acid	Bentonite and kaolin	–	[83]

and the removal of organic contaminants (e.g., dyes, pesticides, and pharmaceuticals/drugs) and common industrial organic wastes (e.g., phenols and aromatic amines) from aqueous solutions occurred by natural adsorbents as shown in Table 2. During the disintegration procedure of organic pollutants, it was observed that the dissolved oxygen in the receiving water frenzied at a greater rate than replenished. It caused oxygen depletion and has harsh consequences for the river biota. And the other side effects occurred when wastewater with organic pollutants contains great quantities of overhanging solids which decrease the light availability to photosynthetic organisms [64].

Adsorption is an exterior fact, with a common method for organic pollutant removal from aqueous solution. It was observed in these techniques that a solution containing absorbable solute when comes to getting in touch with a solid with a very porous surface arrangement, then it was found that liquid–solid intermolecular forces of attraction caused some of the solute molecules from the solution to be concentrated or deposited at the surface. It was studied that clays which were attractive and inexpensive material for the removal of organic contaminants from aqueous solution [65]. There are lots of organic contaminants like pesticides, phenols, and chlorophenols which were adsorbed in water have been studied recently [66–71], and other organic contaminants like chloroacetic acids, such as trichloroacetic acid (TCAA), dichloroacetic acid (DCAA), and monochloroacetic acid (MCAA) are also growing alertness in the literature [72, 73]. A bentonite-based adsorptive and quaternary ammonium salts a modified bentonite adsorbent were used for the removal of organic contaminants dichloroacetic acid (DCAA) and tetrachloride (CT) from water [74, 75]. Inorganic–organic-intercalated (IO) bentonites were modified with Co^{2+}, Ni^{2+}, or Cu^{2+} to create adsorbents for the removal of relevant emerging contaminants (naproxen, salicylic acid, clofibric acid, and carbamazepine) from water, overcoming challenges associated with low concentration and polar nature of these contaminants by relying on weak chemical complexation interactions [76]. For the removal of phenol from aqueous solutions, [77] a natural bentonite modified with a cationic surfactant, cetyl trimethyl ammonium bromide (CTAB) was studied as an adsorbent. For the removal of humic acid (HA) and o-dichlorobenzene (DCB) from drinking water, a combined coagulation process (COBC) was investigated [78, 79]. It was investigated that aluminum sulfate (AS) and poly-aluminum chloride (PACl) used for the removal of the organic composite from the water. It was studied that for the removal of harmful algae from the water, a combination of montmorillonite-Cu (II)/Fe (III) oxides was prepared [80].

The removal of atrazine (2-chloro-4-ethylamino-6-isopropylamino-s-triazine) has been widely used for the removal from water [81]. It was studied that the vesicle–clay [14] complex which was another absorbing material for organic contaminants from water by filtration and sedimentation [82]. It was studied [83] that the two modified clays: bentonite and kaolin were studied for the removal of organic contaminants from aqueous solution.

It was studied that liquefied gasses used for the removal of the organic component from aqueous solution [84]. It was studied that activated carbons (AC) (both granular activated carbon (GAC) and powdered activated carbons (PAC)) are common adsorbents used for the removal of organic contaminants [85, 86]. Cellulose acetate (CA) embedded with triolein (CA-triolein) was prepared as an adsorbent for the removal of persistent organic pollutants (POPs) from micro-polluted aqueous solution [87]. Natural adsorbents include charcoal, clays, clay minerals, zeolites, and ores. These natural materials, in many instances, are relatively cheap, abundant in supply and have significant potential for modification and ultimately enhancement of their adsorption capabilities. Synthetic adsorbents are adsorbents prepared from agricultural products and wastes, household wastes, industrial wastes, sewage sludge, and polymeric adsorbents. Each adsorbent has its own characteristics such as porosity, pore structure, and nature of its adsorbing surfaces [88].

2.3 Pathogens

Pathogens, microscopic biological organisms capable of causing disease, include viruses (comprising DNA or RNA with a protein coating), bacteria (single-celled organisms), protozoa (also single-celled, but with a distinct membrane-bound nucleus), and toxins released by algae (aquatic photosynthetic unicellular or multi-cellular species).

The harmful effects of pathogens range from mild acute illness, through chronic severe sickness to fatality. Important waterborne (transmission via consumption of contaminated water), water-washed (where the quality of used cleansing water is of lesser consideration itself acts as a pathogen source) and water-based (the pathogen or an intermediate host spends part of its life cycle in water) diseases kill millions annually. The World Health Organization (WHO) states 2.16 million people died of diarrheal diseases globally in 2004, more than 80% of which were from low-income countries. Cholera, giardiasis, infectious hepatitis, typhoid, amebic and bacillary dysenteries, and bilharzia are some of the more common diseases responsible.

By far, the most common transmission route is the oral consumption route of pathogens, derived from human feces, or urine residing in contaminated water, including cleaning/washing water. Although many pathogens can live for only a short time outside the human body, waterborne transmission of resilient bacterial cysts and oocysts, together with direct pathogen transport, is a key infection mechanism. Animal feces and urine also harbor important pathogenic species (e.g., leptospirosis) while further risks, particularly to livestock, are derived from the excretion of toxins to water by algae and other microbes (e.g., cyanobacteria). In recent years, potential pathogen contamination from increased land-spread sewage sludge has required greater attention to sludge treatment to alleviate the risk.

Three categories of biological contaminants namely microorganisms, natural organic matter (NOM), and biological toxins have been studied [89–91]. The removal of cyanobacterial toxins from water is a concern in conventional water treatment [92, 93]. Many adsorbents like activated carbon have studied for the good removal of pathogens from water, and lot of factors control the removal process [94]. Different types of nanomaterials such as Ag, titanium, and zinc material have been used for the removal of the pathogen from the water. It was studied that low-cost filter materials coated with silver nanoparticles used for the removal of pathogens from water. It was studied that [95–97] for the removal of pathogens like microcystin cyanobacterial hepatotoxins, natural clay was used. There are a lot of methods using for water treatment [98–105]. Table 3 shows the application of material for the removal of the pathogen from the water

Table 3 Appliance of natural clay and its composites in removing pathogens contaminants from drinking water

Contaminants	Type of clay and modification	Efficiency (%)	Refs.
Microcystin-LR	Natural clay minerals consisting of kaolin and montmorillonite	81	[89]
Cryptosporidium	Natural clay minerals consisting of kaolin and montmorillonite	72	[90]
Giardia lamblia	Natural clay minerals consisting of kaolin and montmorillonite	75	[91]
Legionella	Natural clay minerals consisting of kaolin and montmorillonite	80	[92]
Escherichia coli	AgNPs	–	[95]

3 Summary

All tables verified that the multiplicity of pollutants treated with different types of clays and has been removed from the water. Now, it was observed from the tables that natural clay and its composites have the capability for the removal of contaminants ranging from metals to pollutants. It was observed from the results that by using natural clay and its modified composites, they showed eco-friendly nature. They were capable of removing organic and inorganic contaminants from drinking water with the very high removal of toxic trace metals, nutrients, and organic matter. After the research, the superiority of adsorbents has been proven and it was found that the water treatment adsorbent can be reused. For decontamination of drinking water, the conventional method has been used. Because of natural and their plenty presence of adsorbent made them a low-cost material and harmless adsorbent which could be used for removal of different contaminants from water. It has been helpful for developing nations. Natural treatments have been eco-friendly, best removal efficiency, and low-cost methods. A lot of bio-adsorbents have been easily available in nature. There are lots of chemicals in the form of inorganic or organic coming out from industries, so different kind of natural material has been used for the removal of this kind of contaminants from water.

4 Future Scenario

The natural materials with nanomaterial covering hold great assure for water treatment. As discussed above, the adsorption ability of natural and modified minerals increases with the covering of nanomaterial on them. In this field, more research is essential to get good results in using the hybrid materials for water treatment. Further, research which needs instant concentration involves using a natural material with nanomaterials for flourishing removal of rising contaminants present in large amount in our drinking water. I have studied that present water treatment methods are ineffectual of removing the rising contaminants. At present, the research in this field is limited. But the existing study results hold important promise for the use of modified materials for rising contaminant treatment without undesired poisonous effects to the ecosystem.

Acknowledgements We gratefully acknowledge support from the Ministry of Human Resource Development, Department of Higher Education, Government of India under the scheme of Establishment of Centre of Excellence for Training and Research in Frontier Areas of Science and Technology (FAST), vide letter No, F. No. 5-5/201 4-TS. Vll.

References

1. The Clay Mineral Group. 2011. http://mineral.galleries.com/minerals/silicate/clays.htm.
2. Lin, S.H., and R.S. Juang. 2002. Heavy metal removal from water by sorption using surfactant-modified montmorillonite. *Journal of Hazardous Materials* 92 (3): 315–326.
3. Krishna, B.S., D.S.R. Murty, and B.S. Jai Prakash. 2000. Thermodynamics of chromium(VI) anionic species sorption onto surfactant-modified montmorillonite clay. *Journal of Colloid and Interface Science* 229 (1): 230–236.
4. Bailey, S.E., T.J. Olin, R.M. Bricka, and D.D. Adrian. 1999. A review of potentially low-cost sorbents for heavy metals. *Water Research* 33 (11): 2469–2479.
5. Babel, S., and T.A. Kurniawan. 2003. Low-cost adsorbents for heavy metals uptake from contaminated water: A review. *Journal of Hazardous Materials* 97 (1–3): 219–243.
6. Virta, R.L. 1996. *U.S. Geological Survey-Minerals Information*, http://minerals.usgs.gov/minerals/pubs/commodity/190496.pdf.
7. Pinnavaia, T.J. 1983. Intercalated clay catalysts. *Science* 220 (4595): 365–371.
8. Cadena, F. Rizvi, R. and Peters, R. W. (1990). Feasibility studies for the removal of heavy metal from solution using tailored bentonite, hazardous and industrial wastes. In *Proceedings of the 22nd Mid-Atlantic Industrial Waste Conference*, Drexel University, 77–94.
9. Tanabe, K. 1981. Solid acid and base catalysis. In *Catalysis—Science and technology*, edited by J.R. Anderson and M. Boudart, 231.
10. Olphen, H. 1977. *An introduction to clay colloid chemistry*. New York, NY, USA: Wiley-Interscience.
11. Churchman, G.J. 2002. Formation of complexes between bentonite and different cationic polyelectrolytes and their use as sorbents for non-ionic and anionic pollutants. *Applied Clay Science* 21 (3–4): 177–189.
12. Breen, C. 1999. The characterisation and use of polycationexchanged bentonites. *Applied Clay Science* 15 (1–2): 187–219.
13. Radian, A., and Y.G. Mishael. 2008. Characterizing and designing polycation—clay nanocomposites as a basis for imazapyr controlled release formulations. *Environmental Science and Technology* 42 (5): 1511–1516.
14. Zadaka, D., S. Nir, A. Radian, and Y.G. Mishael. 2009. Atrazine removal from water by polycation-clay composites Effect of dissolved organic matter and comparison to activated carbon. *Water Research* 43 (3): 677–683.
15. Darder, M., M. Colilla, and E. Ruiz-Hitzky. 2005. Chitosan-clay nanocomposites: Application as electrochemical sensors. *Applied Clay Science* 28 (1–4): 199–208.
16. Darder, M., M.L. Blanco, P. Aranda, A.J. Aznar, J. Bravo, and E. Ruiz-Hitzky. 2006. Microfibrous chitosan—sepiolite nanocomposites. *Chemistry of Materials* 18 (6): 1602–1610.

17. Ruiz-Hitzky, E., M. Darder, and P. Aranda. 2005. Functional biopolymer nanocomposites based on layered solids. *Journal of Materials Chemistry* 15 (35–36): 3650–3662.
18. An, J.H., and S. Dultz. 2007. Adsorption of tannic acid on chitosan montmorillonite as a function of pH and surface charge properties. *Applied Clay Science* 36 (4): 256–264.
19. Li, J.M., X.G. Meng, C.W. Hu, and J. Du. 2009. Adsorption of phenol, p-chlorophenol, and p-nitrophenol onto functional chitosan. *Bioresource Technology* 100 (3): 1168–1173.
20. Bhattacharyya, K.G., and S.S. Gupta. 2008. Adsorption of a few heavy metals on natural and modified kaolinite and montmorillonite: A review. *Advances in Colloid and Interface Science* 140 (2): 114–131.
21. Ulmanu, M., E. Marañón, Y. Fernández, L. Castrillón, I. Anger, and D. Dumitriu. 2003. Removal of copper and cadmium ions from diluted aqueous solutions by low cost and waste material adsorbents. *Water, Air, and Soil pollution* 142 (1–4): 357–373.
22. Yavuz, O., Y. Altunkaynak, and F. Guzel. 2003. Removal of copper, nickel, cobalt and manganese from aqueous solution by kaolinite. *Water Research* 37 (4): 948–952.
23. Bhattacharyya, K.G., and S.S. Gupta. 2007. Adsorption of Co(II) from aqueous medium on natural and acid activated kaolinite and montmorillonite. *Separation Science and Technology* 42 (15): 3391–3418.
24. Bhattacharyya, K.G., and S.S. Gupta. 2006. Adsorption of Fe(III) from water by natural and acid activated clays: Studies on equilibrium isotherm, kinetics, and thermodynamics of interactions. *Adsorption* 12 (3): 185–204.
25. Gupta, S.S., and K.G. Bhattacharyya. 2005. Interaction of metal ions with clays: I. A case study with Pb(II). *Applied Clay Science* 30 (3–4): 199–206.
26. Mellah, A., and S. Chegrouche. 1997. The removal of zinc from aqueous solutions by natural bentonite. *Water Research* 31 (3): 621–629.
27. Oliveira, L.C.A., R.V.R.A. Rios, J.D. Fabris, K. Sapag, V.K. Garg, and R.M. Lago. 2003. Clay-iron oxide magnetic composites for the adsorption of contaminants in water. *Applied Clay Science* 22 (4): 169–177.
28. Etci, Ö., N. Bektaş, and M.S. Öncel. 2010. Single and binary adsorption of lead and cadmium ions from aqueous solution using the clay mineral beidellite. *Environmental Earth Sciences* 61 (2): 231–240.
29. Gecol, H., P. Miakatsindila, E. Ergican, and R.H. Sage. 2006. Biopolymer coated clay particles for the adsorption of tungsten from water. *Desalination* 197 (1–3): 165–178.
30. Aytas, S., M. Yurtlu, and R. Donat. 2009. Adsorption characteristic of U(VI) ion onto thermally activated bentonite. *Journal of Hazardous Materials* 172 (2–3): 667–674.
31. Mishra, P.C., and R.K. Patel. 2009. Removal of lead and zinc ions from water by low-cost adsorbents. *Journal of Hazardous Materials* 168 (1): 319–325.
32. Yuan, P., M. Fan, and D. Yang. 2009. Montmorillonite-supported magnetite nanoparticles for the removal of hexavalent chromium [Cr(VI)] from aqueous solutions. *Journal of Hazardous Materials* 166 (2–3): 821–829.
33. Angove, M.J., B.B. Johnson, and J.D. Wells. 1998. The influence of temperature on the adsorption of cadmium(II) and cobalt(II) on kaolinite. *Journal of Colloid and Interface Science* 204 (1): 93–103.
34. Doušová, B., L. Fuitová, and T. Grygar. 2009. Modified aluminosilicates as low-cost sorbents of As(III) from anoxic groundwater. *Journal of Hazardous Materials* 165 (1–3): 134–140.
35. Sajidu, S.M.I., I. Persson, W.R.L. Masamba, E.M.T. Henry, and D. Kayambazinthu. 2006. Removal of Cd^{2+}, Cr^{3+}, Cu^{2+}, Hg^{2+}, Pb^{2+} and Zn^{2+} cations and AsO_3^{-4} anions from aqueous solutions by mixed clay from Tundulu in Malawi and characterisation of the clay. *Water SA* 32 (4): 519–526.
36. Bleiman, N., and Y.G. Mishael. 2010. Selenium removal from drinking water by adsorption to chitosan-clay composites and oxides: Batch and columns tests. *Journal of Hazardous Materials* 183 (1–3): 590–595.
37. Na, P., X. Jia, and B. Yuan. 2010. Arsenic adsorption on Ti-pillared montmorillonite. *Journal of Chemical Technology and Biotechnology* 85 (5): 708–714.

38. Bulut, Y., G. Akçay, D. Elma, and I.E. Serhatlı. 2009. Synthesis of clay-based superabsorbent composite and its sorption capability. *Journal of Hazardous Materials* 171 (1–3): 717–723.
39. Chaturvedi, A.K., K.P. Yadava, K.C. Pathak, and V.N. Singh. 1990. Defluoridation of water by adsorption on fly ash. *Water, Air, and Soil Pollution* 49 (1–2): 41–69.
40. Sujana, M.G., R.S. Thakur, and S.B. Rao. 1998. Removal of fluoride from aqueous solution by using alum sludge. *Journal of Colloid and Interface Science* 206 (1): 94–101.
41. Toyoda, A., and T. Taira. 2000. A new method for treating fluorine wastewater to reduce sludge and running costs. *IEEE Transactions on Semiconductor Manufacturing* 13 (3): 305–309.
42. Ayoob, S., and A.K. Gupta. 2006. Fluoride in drinking water: A review on the status and stress effects. *Critical Reviews in Environmental Science and Technology* 36 (6): 433–487.
43. WHO (World Health Organization). 1984. *Fluorine and fluorides*. Geneva, Switzerland, World Health Organization: Environmental Health Criteria.
44. Thakre, D., S. Rayalu, R. Kawade, S. Meshram, J. Subrt, and N. Labhsetwar. 2010. Magnesium incorporated bentonite clay for defluoridation of drinking water. *Journal of Hazardous Materials* 180 (1–3): 122–130.
45. Kamble, S.P., P. Dixit, S.S. Rayalu, and N.K. Labhsetwar. 2009. Defluoridation of drinking water using chemically modified bentonite clay. *Desalination* 249 (2): 687–693.
46. Ma, Y.X., F.M. Shi, X.L. Zheng, J. Ma, and J.M. Yuan. 2005. Defluoridation from aqueous solutions by Zr-loaded bentonite. *Journal of Harbin Institute of Technology (New Series)* 12 (1): 224–229.
47. Dhillon, A., and D. Kumar. 2015. Development of a nanoporous adsorbent for the removal of health-hazardous fluoride ions from aqueous systems. *Journal of Material Chemistry A* 3: 4215–4228.
48. Dhillon, A., and D. Kumar. 2015. Nanocomposite for the detoxification of drinking water: Effective removal of fluoride and bactericidal activity. *New Journal of Chemistry* 39: 9143–9154.
49. Tomar, V., S. Prasad, and D. Kumar. 2013. Adsorptive removal of fluoride from water samples using Zr-Mn composite material. *Microchemical Journal* 111: 116–124.
50. Bejaoui, I., A. Mnif, and B. Hamrouni. 2014. Performance of reverse osmosis and nanofiltration in the removal of fluoride from model water and metal packaging industrial effluent. *Separation Science and Technology* 49: 1135–1145.
51. Kotecha, P.V., S.V. Patel, K.D. Bhalani, D. Shah, V.S. Shah, and K.G. Mehta. 2012. Prevalence of dental fluorosis & dental caries in association with high levels of drinking water fluoride content in a District of Gujarat, India. Development Foundation, New Delhi. *Indian Journal of Medical Research* 135: 873–877.
52. Cui, H., Y. Qian, H. An, C. Sun, J. Zhai, and Q. Li. 2012. Electrochemical removal of fluoride from water by PAOA modified carbon felt electrodes in a continuous flow reactor. *Water Research* 46: 3943–3950.
53. Guo, Q., and E.J. Reardon. 2012. Fluoride removal from water by meixnerite and its calcination product. *Applied Clay Science* 56: 7–15.
54. Ramanjaneyulu, V., M. Jaipal, N. Yasovardhan, and S. Sharada. 2013. Kinetic studies on removal of fluoride from drinking water by using tamarind shell and pipal leaf powder. *International Journal of Emerging Trends in Engineering and Development* 5: 146.
55. Sakhare, N., S. Lunge, R. Rayalu, S. Bakardjiva, J. Subrt, S. Devotta, and N. Labhsetwar. 2012. Defluoridation of water using calcium aluminate material. *Chemical Engineering Journal* 203: 406–414.
56. Chakrabarty, S., and H.P. Sarma. 2012. Defluoridation of contaminated drinking water using neem charcoal adsorbent: Kinetics and equilibrium studies. *International Journal of Chem Tech Research* 4: 511–516.
57. Boubakri, A., N. Helali, M. Tlili, and M.B. Amor. 2014. Fluoride removal from diluted solutions by Donnan dialysis using full factorial design. *Korean Journal of Chemical Engineering* 31 (3): 461–466.

58. Babu, J.M., and S. Goel. 2013. Defluoridation of drinking water in batch and continuous-flow electrocoagulation systems. *Pollution Research* 32 (4): 727–736.
59. Andey, S., P.K. Labhasetwar, G. Khadse, P. Gwala, P. Pal, and P. Deshmukh. 2013. Performance evaluation of solar power based electrolytic defluoridation plants in India. *International Journal of Water Resources and Arid Environments* 2 (3): 139–145.
60. Takdastan, A., S.E. Tabar, A. Neisi, and A. Eslami. 2014. Fluoride removal from drinking water by electrocoagulation using iron and aluminum electrodes, Jundishapur. *Journal of Health Science* 6 (3): 39–44.
61. Sandoval, M.A., R. Fuentes, J.L. Nava, and I. Rodríguez. 2014. Fluoride removal from drinking water by electrocoagulation in a continuous filter-press reactor coupled to a flocculation and clarifier. *Separation and Purification Technology* 134: 163–170.
62. Naim, M. M. Moneer, A. A., and El-Said, G. F. 2015. Predictive equations for the defluoridation by electrocoagulation technique using bipolar aluminum electrodes in the absence and presence of additives: A multivariate study. *Desalination and Water Treatment*, 1–13.
63. Mena-Duran, C.J., M.R. Sun Kou, and T. Lopez. 2007. Nitrate removal using natural clays modified by acid thermoactivation. *Applied Surface Science* 253 (13): 5762–5766.
64. Murray, H.H. 2000. Traditional and new applications for kaolin, smectite, and palygorskite: A general overview. *Applied Clay Science* 17 (5–6): 207–221.
65. Camazano, M.S., and M.J.S. Martin. 1983. Factors influencing interactions of organophosphorus pesticides with montmorillonite. *Geoderma* 29 (2): 107–118.
66. Ainsworth, C.C., J.M. Zachara, and R.L. Schmidt. 1987. Quinoline sorption on Na-montmorillonite: contributions of the protonated and neutral species. *Clays and Clay Minerals* 35 (2): 121–128.
67. Khoshnood, M., and S. Azizian. 2012. Adsorption of 2,4-dichlorophenoxyacetic acid pesticide by graphitic carbon nanostructures prepared from biomasses. *Journal of Industrial and Engineering Chemistry* 18 (5): 1796–1800.
68. Rodriguez, J.M., A.J. Lopez, and S. Bruque. 1988. Interaction of phenamiphos with montmorillonite. *Clays & Clay Minerals* 36 (3): 284–288.
69. Shu, H.T., D. Li, A.A. Scala, and Y.H. Ma. 1997. Adsorption of small organic pollutants from aqueous streams by aluminosilicate-based microporous materials. *Separation and Purification Technology* 11 (1): 27–36.
70. Torrents, A., and S. Jayasundera. 1997. The sorption of nonionic pesticides onto clays and the influence of natural organic carbon. *Chemosphere* 35 (7): 1549–1565.
71. Danis, T.G., T.A. Albanis, D.E. Petrakis, and P.J. Pomonis. 1998. Removal of chlorinated phenols from aqueous solutions by adsorption on alumina pillared clays and mesoporous alumina aluminum phosphates. *Water Research* 32 (2): 295–302.
72. Konstantinou, I.K., T.A. Albanis, D.E. Petrakis, and P.J. Pomonis. 2000. Removal of herbicides from aqueous solutions by adsorption on Al-pillared clays, Fe-Al pillared clays, and mesoporous alumina aluminum phosphates. *Water Research* 34 (12): 3123–3136.
73. Sun, D., W. Cai, C. Shi, X. Mu, Y. Song, and H. Qi. 2000. Advanced oxidations of chloroacetic acids present in drinking water. *Journal of Environmental Science and Health A* 35 (10): 1811–1816.
74. Pervova, M.G., V.E. Kirichenko, and K.I. Pashkevich. 2002. Determination of chloroacetic acids in drinking water by reaction gas chromatography. *Journal of Analytical Chemistry* 57 (4): 326–330.
75. Gu, L., X. Yu, J. Xu, L. Lv, and Q. Wang. 2011. Removal of dichloroacetic acid from drinking water by using adsorptive ozonation. *Ecotoxicology* 20 (5): 1160–1166.
76. Lu, J., and Pan, J. 2010. Removal of carbon tetrachloride from contaminated groundwater environment by adsorption method. In *Proceedings of the 4th International Conference on Bioinformatics and Biomedical Engineering (iCBBE'10)* Chengdu, China.
77. Rivera-Jimenez, S.M., M.M. Lehner, W.A. Cabrera-Lafaurie, and A. J. Hernández-Maldonado. 2011. Removal of naproxen, salicylic acid, clofibric acid, and

carbamazepine by water phase adsorption onto inorganic-organic-intercalated bentonites modified with transition metal cations. *Environmental Engineering Science* 28 (3): 171–182.
78. Senturk, H.B., D. Ozdes, A. Gundogdu, C. Duran, and M. Soylak. 2009. Removal of phenol from aqueous solutions by adsorption onto organomodified Tirebolu bentonite: equilibrium, kinetic and thermodynamic study. *Journal of Hazardous Materials* 172 (1): 353–362.
79. Gu, L., X. Zhang, L. Lei, and X. Liu. 2009. Concurrent removal of humic acid and o-dichlorobenzene in drinking water by combined ozonation and bentonite coagulation process. *Water Science and Technology* 60 (12): 3061–3068.
80. Jiang, J.Q., and C.G. Kim. 2008. Comparison of algal removal by coagulation with clays and Al-based coagulants. *Separation Science and Technology* 43 (7): 1677–1686.
81. Gao, Z., X. Peng, H. Zhang, Z. Luan, and B. Fan. 2013. Montmorillonite-Cu(II)/Fe(III) oxides magnetic material for removal of cyanobacterial Microcystis aeruginosa and its regeneration. *Desalination* 247 (1–3): 337–345.
82. Undabeytia, T., S. Nir, J. Sánchez-Verdejo, J. Villaverde, C. Maqueda, and E. Morillo. 2008. A clay-vesicle system for water purification from organic pollutants. *Water Research* 42 (4–5): 1211–1219.
83. Rytwo, G., Y. Kohavi, I. Botnick, and Y. Gonen. 2007. Use of CV and TPP-montmorillonite for the removal of priority pollutants from water. *Applied Clay Science* 36 (1–3): 182–190.
84. Bonina, F.P., M.L. Giannrossi, L. Medici, C. Puglia, V. Summa, and F. Tateo. 2007. Adsorption of salicylic acid on bentonite and kaolin and release experiments. *Applied Clay Science* 36 (1–3): 77–85.
85. Wang, T., R.L. Zhu, F. Ge, J.X. Zhu, H.P. He, and W.X. Chen. 2010. Sorption of phenol and nitrobenzene in water by CTMAB/CPAM oregano bentonites. *Huanjing Kexue/Environmental Science* 31 (2): 385–389.
86. Carmichael, W.W. 1988. Freshwater cyanobacteria (blue-green algal) toxins. In *Natural toxins: Characterization, pharmacology and therapeutics*, edited by C.L. Ownby and G.V. Odell, 3–16. London, UK: Pergamon Press.
87. Cohen, P., and P.T.W. Cohen. 1989. Protein phosphatases come of age. *Journal of Biological Chemistry* 264 (36): 21435–21438.
88. Yoshizawa, S., R. Matsushima, and M.F. Watanabe. 1990. Inhibition of protein phosphatases by microcystis and nodularin associated with hepatotoxicity. *Journal of Cancer Research and Clinical Oncology* 116 (6): 609–614.
89. Honkanen, R.E., J. Zwiller, and R.E. Moore. 1990. Characterization of microcystin-LR, a potent inhibitor of type 1 and type 2A protein phosphatases. *Journal of Biological Chemistry* 265 (32): 19401–19404.
90. MacKintosh, C., K.A. Beattie, S. Klumpp, P. Cohen, and G.A. Codd. 1990. Cyanobacterial microcystin-LR is a potent and specific inhibitor of protein phosphatases 1 and 2A from both mammals and higher plants. *FEBS Letters* 264 (2): 187–192.
91. Nishiwaki-Matsushima, R., S. Nishiwaki, and T. Ohta. 1991. Structure- function relationships of microcystins, liver tumor promoters, in interaction with protein phosphatase. *Japanese Journal of Cancer Research* 82 (9): 993–996.
92. Nishiwaki-Matsushima, R., T. Ohta, and S. Nishiwaki. 1992. Liver tumor promotion by the cyanobacterial cyclic peptide toxin microcystin-LR. *Journal of Cancer Research and Clinical Oncology* 118 (6): 420–424.
93. Fujiki, H., and M. Suganuma. 1993. Tumor promotion by inhibitors of protein phosphatases 1 and 2A: The okadaic acid class of compounds. *Advances in Cancer Research* 61: 143–194.
94. Lawton, L.A., B.J.P.A. Cornish, and A.W.R. MacDonald. 1998. Removal of cyanobacterial toxins (microcystins) and cyanobacterial cells from drinking water using domestic water filters. *Water Research* 32 (3): 633–638.
95. Heidarpour, F., and Wan. W. 2011. Complete removal of pathogenic bacteria from water using nano silver coate cylindrical polypropylene. *Journal of Toxicology—Toxin Reviews* 17 (3):385–403.
96. Xagoraraki, I., Yin, Z., and Svambayev, Z. (2014). Fate of viruses in water systems. *Journal of Environment Engineering*, 140. doi:10.1061/(ASCE)EE.1943-7870.0000827.

97. Lu, R., D. Mosiman, and T.H. Nguyen. 2013. Mechanisms of MS2 bacteriophage removal by fouled ultrafiltration membrane subjected to different cleaning methods. *Environmental Science and Technology* 47: 13422–13429.
98. Antony, A., J. Blackbeard, and G. Leslie. 2011. Removal efficiency and integrity monitoring techniques for virus removal by membrane processes. *Critical Reviews in Environmental Science and Technology* 42: 891–933.
99. Hirani, Z.M., Z. Bukhari, J. Oppenheimer, P. Jjemba, M.W. LeChevallier, and J.G. Jacangelo. 2014. Impact of MBR cleaning and breaching on passage of selected microorganisms and subsequent inactivation by free chlorine. *Water Research* 57: 313–324.
100. Cabral, J.P.S. 2010. Water microbiology. Bacterial pathogens and water. *International Journal of Environmental Research and Public Health* 7: 3657–3703.
101. Luo, W., F.I. Hai, W.E. Price, W. Guo, H.H. Ngo, K. Yamamoto, and L.D. Nghiem. 2014. High retention membrane bioreactors: Challenges and opportunities. *Bioresources Technology* 167: 539–546.
102. Amin, M.T., A.A. Alazba, and U. Manzoor. 2014. A review of removal of pollutants from water/wastewater using different types of nanomaterials. *Advances in Materials Science and Engineering* 23 (4): 23–28.
103. Botes, M., and T. Eugene Cloete. 2010. The potential of nanofibers and nano biocides in water purification. *Critical Reviews in Microbiology* 36: 68–81.
104. Homaeigohar, S., and M. Elbahri. 2014. Nanocomposite electrospun nanofiber membranes for environmental remediation. *Materials* 7: 1017–1045.
105. Semblante, G.U., F.I. Hai, H.H. Ngo, W. Guo, S.J. You, W.E. Price, and L.D. Nghiem. 2014. Sludge cycling between aerobic, anoxic and anaerobic regimes to reduce sludge production during wastewater treatment: Performance, mechanisms, and implications. *Bioresource Technology* 155: 395–409.

Synthesis and Application of Silica Nanoparticles-Based Biohybrid Sorbents

Ritu Painuli, Sapna Raghav and Dinesh Kumar

Abstract Progress in the silicon oxide/polymer hybrid composite materials combines the unique attributes of the inorganic fillers and the organic polymers. Organic/inorganic nanocomposites are usually consulted as the organic polymer having a building block of the inorganic nanoscale. To recognize the interface interaction, nanoscale hybridization of organic polymers and Silica fillers, a new approach has been worked to synthesize hybrid materials in nanotechnology. Thus, this chapter explores the preparation Silica hybrid composite materials and their broad applications, such as functional coatings, biomedical applications, and so on.

Keywords Silica nanoparticles · Biohybrids · Mesoporous Silica · Preparation · Applications

1 Introduction

In the application fields over the last few decades, traditional polymers have been replaced by organic polymer–inorganic oxide filler composites. Organic/inorganic nanohybrid composites are usually organic polymer composites with inorganic nanoscale building units. These composites show inimitable properties combining the advantages of the inorganic fillers like the rigidity, enhanced thermal stability, mechanical property with the processability, flexibility, and ductility of the organic polymers [1]. In comparison to the conventional composites, a significant characteristic of polymer nanocomposite with nano-size inorganic filler is the spectacular increase in the interfacial area. The properties of the composites even at low loadings are essentially unique in relation to that of the bulk polymer. There are various inorganic nano-sized building blocks, for instances, carbon nanotubes,

---

R. Painuli · S. Raghav
Department of Chemistry, Banasthali University, Vanasthali 304022, Rajasthan India

D. Kumar (✉)
School of Chemical Science, Central University of Gujarat, Gandhinagar 382030, India
e-mail: dinesh.kumar@cug.ac.in

layered silicates, i.e., saponite, metal nanoparticles and metal oxide, spherical Silica particles, etc. Amid all these inorganic oxide fillers, Silica particles have gained much more attractive and have been utilized in a variety of applications. This is because of their unique properties such as economic production, well-defined structure, enhanced surface area as well as the effortless surface modification [2–5]. During the composition of polymer with Silica (Si), the SiNPs can enhance the thermal properties along with the self-sustaining capability under required conditions. On comparing with unmodified polymers, the Silica nanoparticles modification leads to the improved hydrophobic interactions with the polymer [6]. The synthesis, characterization, properties, and applications of Silica nanohybrid composites have become a rapidly increasing topic in the research area. There is various literature available in the field of Silica/polymer nanohybrid composites, for instances, Althues et al. in 2007 [7] reported the incorporation of functional inorganic nanofillers onto the polymeric materials, the sol–gel method for the Silica-based materials and recent applications [3]. Thus, the present chapter explores the current trends in the progress of Silica (Si)/polymer hybrid composites: materials and applications, for instances, metal uptake and bio applications.

2 Silica Amendments

Initially from the Silica precursors, the Silica particles which are synthesized by methods such as thermal and wet routes, and from the plants extract [8–10], have numerous OH groups known as silanols. These silanols are having reactive sites and are hydrophilic in nature and can be changed into another valuable functionality. The functional groups present on the Silica particles surface play a very fundamental role in the definite properties of composites materials, for instances, hydrophobicity, hydrophilicity, chemical binding capacity, etc. [11, 12]. The amendment normally includes modifying the properties of the surface of oxide fillers. By the assistance of the sol–gel process, nanofillers, for instances, porous and spherical Silica particles could be obtained. When the Silica fillers are evenly dispersed in the polymer matrix, then the excellent efficiency of polymer composites could be accomplished. The advanced properties are generated during the preparation of the homogeneous mixtures. On modifying the hydrophilic surface of Silica fillers, the hydrophobic property gets changed into hydrophobic and reactive. This can be done by a physical method (by physisorption) or by chemical method via the covalent bonding [13]. However, owing to the comparatively weak H-bonding or van der Waals forces among the two phases, the physisorbed modification method suffers from severe drawbacks [14]. On the other hand, the formation of the organic polymer and the Silica filler can be achieved by the condensation reaction in the sol–gel method. As a result, a chemical bond is formed between the Silica filler and the organic polymer. Thus, to improve the interfacial stability between the Silica fillers and the organic polymer, these covalent grafting

2.1 Physical Interaction

Based upon the physical interaction, the amendment of the Silica fillers is generally executed by utilizing the macromolecules or surfactants absorbed over the Silica nanoparticles surface [16]. The favored adsorption of a polar group of the macromolecule onto the surface of the Silica fillers via the electrostatic interaction or hydrogen bonding is the chief principle of the macromolecule treatment [17]. By decreasing the physical interaction, a macromolecule can effortlessly decrease the interactions among the Silica particles in the agglomerates. A surfactant can reduce the interaction between the Silica nanofillers through reducing the physical attraction and can easily be incorporated into a polymer matrix. For instance, to increase the disparity and the interaction between the Silica fillers and the organic polymers, the Silica NPs were treated with stearic acid [18] and oleic acid [19–21] and with a cationic salt such as CTAB [22, 23]. A modification was done on the Si surface with the adsorption of an oxyethylene-based macromonomer by Reculusa et al. [24]. Thus, the macromonomer is mainly hydrophilic. These are then capable of forming H-bonding with the silanol functions present on the Si surface.

2.2 Chemical Interaction

The mainly utilized method for the composition of the organic polymers is the silylation of the –OH groups on the Silica fillers by utilizing organosilane coupling agents [25, 26]. Therefore, it is the amendment of the inorganic oxide particles via the chemical reaction. Therefore, this is the modification of the inorganic oxide particles by the chemical reaction since it can lead to the highly stronger interaction between the modifiers and the Silica fillers. They possess hydrolysable groups (X) and a non-hydrolyzable organic group (R–Y) on the silicon atom (Fig. 1). Here the non-hydrolyzable organic functional groups which are adhered onto the Silica filler surface act as hydrophobic or reactive sites [27]. The factors which affect the surface treatment of Silica fillers are the type of modifier, time, treatment concentration, and predispose method [27]. Since these factors might interact with each

Fig. 1 A method for the preparation of organosilanols

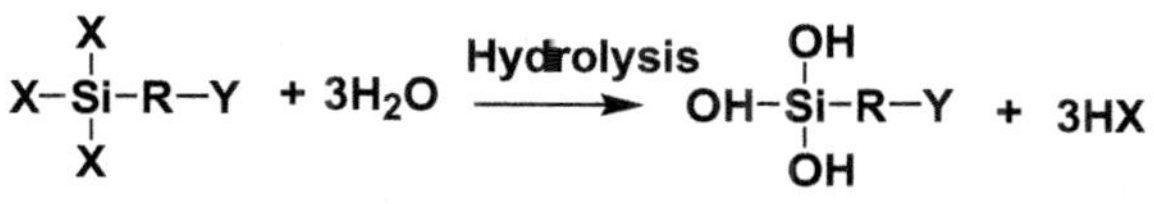

$$X-\underset{X}{\overset{X}{Si}}-R-Y + 3H_2O \xrightarrow{\text{Hydrolysis}} OH-\underset{OH}{\overset{OH}{Si}}-R-Y + 3HX$$

X: Alkoxy, Halide
R; Alkylene, Arylene, Polymeric units
Y; H, NH_2, SH, Alkenyl, Acryloxy, etc.

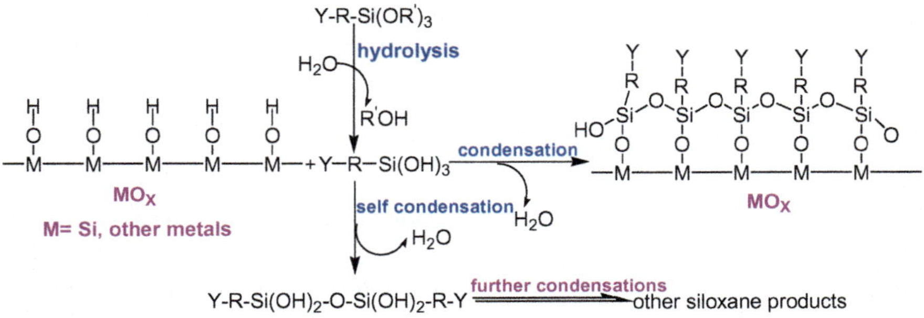

Fig. 2 Modification of the inorganic oxide particles

other in determining the properties of the surface. During the surface amendment of Silica to obtain hydrophobic and reactive fillers, coupling agents such as alkoxysilane are generally utilized in comparison with chlorosilanes due to the production of the byproduct, i.e., hydrochloric acid, from the chlorosilane hydrolysis. Hence, using alkoxysilane as a coupling agent, in the first step coupling agent requires moisture producing reactive organosilaneols (RSi–OH). Then, the produced organosilanol undergo condensation with the hydroxyl groups which are present on the inorganic oxide filler surface to give organofunctional groups (Fig. 2) [28]. A few examples of organofunctional groups are sulfide, thiols, isocyanate, acryloxy, terminal olefinyl groups, etc. The silanetriols can be isolated by the hydrolysis alkoxysilane coupling agents. This can be done under defined conditions and can be used for the amendment of inorganic oxide fillers [28]. During the amendment of Silica, the silanetriols intermediates were utilized as the modifiers. In a similar way, the modification of other inorganic oxides fillers such as alumina, titania, zirconia, tin oxide, and nickel oxide were reported [28]. There are two kinds of trialkoxysilane coupling agent and dialkoxysilane commercially available. Generally trialkoxysilane coupling agents are cheaper and more widely used than dialkoxysilane coupling agents in industrial field.

It was demonstrated by the Bracho et al. that in the composition of polypropylene, Silica nanospheres which are treated with organic chlorosilanes could be utilized as fillers [29]. To prepare the Silica hybrid composites, another effective method is the grafted polymerization on Silica NPs. To attach the chemical polymer chains onto the surface of Silica fillers, the two common methods are (1) binding of the terminal functional polymers onto the surface and (2) in situ polymerization of monomers via the immobilized initiators on Silica. Silica-grafted acrylonitrile-butadiene-styrene copolymer composites were synthesized via an open ring radical reaction of modified Silica, maleic anhydride, and styrene, in ABS tetrahydrofuran solution by Zheng et al. [30]. Synthesis of the polymer brushes grafted on SiNPs was demonstrated by Li et al. [31]. The light irradiation method could also be employed for the preparation of the polymers grafting NPs. On the graft polymerization of Si, there is an increase in the interface interaction because of the molecular entanglement between the grafting polymer on the NPs and in the modification of NPs on the matrix polymer. These polymer/filler composites can also advance some specific properties of the hybrids to fabricate core/shell charged polymer brushes grafted hollow Silica NPs. The synthesized Silica nanoparticles can also be utilized in the drug delivery system [32].

3 Blending Methods

The easiest and conventional method for the preparation of polymer/Silica hybrid composites is the mixing of polymer and Silica. The method via which the mixing is carried out includes solution blending, melt blending, and the in situ polymerization. During these mixing procedures, the major complexity is the effective dispersion of the Silica NPs into the polymer matrix as they usually lead to agglomeration.

3.1 Melt Blending Methods

Owing to its effectual processability, melt blending is the most utilized method [33]. The various advantages related to this method are its simplicity, cost effectiveness, and its applicability. The superiority in the mechanical property of the polymer composites can be obtained by utilizing the injection molding machine and the twin screw extruder. For the composite of the polymer and the Silica, the melt blending is carried out as above the glass-transition temperature (Tg) of the polymer. This way of mixing is operable and proficient in comparison to the effortless mixing at room temperature (Fig. 3) [33]. For the preparation of the nanocomposites of polybutadiene (BR), styrene-butadiene rubber, this procedure is widely adopted, and Fig. 4 represents the coupling between the Silica particles and the unsaturated polymers by utilizing the TESPT agent affording well-hybridized composite materials. For the preparation of the Silica fillers, alternatively, SBR containing 1-methylimidazolium methacrylate as the ionic liquid unit was utilized [34]. Perez et al. demonstrated that acrylonitrile butadiene rubber interacts strongly with the Silica in comparison to the styrene-butadiene rubber. It was analyzed based on the

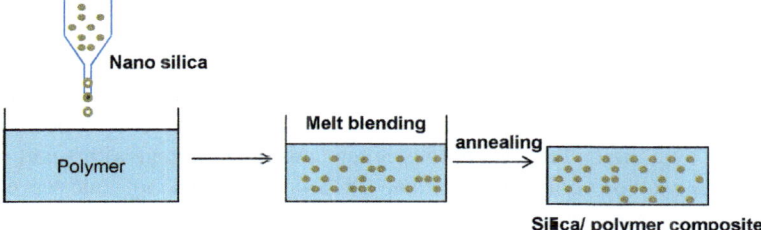

Fig. 3 Melt blending method

Fig. 4 Vulcanization of SBR

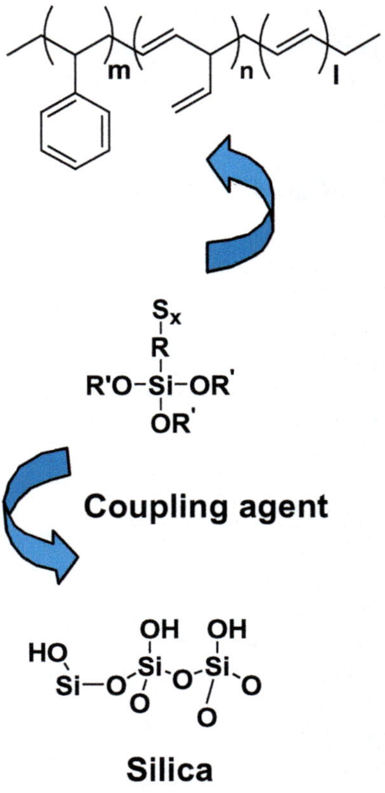

SEM images of the composites of NBR and SBR with mesoporous Silica (MSi) NPs [35]. In comparison to the normal spherical fillers, MSi fillers interact more strongly with the matrices of the polymer. This is owing to the better surface area and porous structure of the MSi fillers. By the penetration of the silicone component into the MSi particles and utilizing using organic solvent by a capillary force, Silicone rubber with low thermal expansion was synthesized [36].

To synthesize epoxy and Silica resin, in the hybrids of MSi particles and fine Silica particles, two types of inorganic fillers are prepared. This preparation can be achieved by the two steps: first, the epoxy resin is mixed mechanically and then the addition of the curing agent.

In comparison with the Silica particles without mesopores, mesoporous Silica particles displayed a better effect on lowering the coefficient of linear thermal expansion values [37]. Suzuki et al. prepared the low dielectric Silica/polymer composites by utilizing the mesoporous Silica particles modified with smart molecular caps, for example, polysilsesquioxane (POSS). The POSS molecule prevents the penetration of the polymers inside the mesopores. Then, to lessen the polarity, the trimethylsilyl agents are grafted onto silanols groups. The prepared composites displayed brilliant low dielectric property [38]. Hwang et al. synthesized organically modified Silica materials of different particle sizes and their polypropylene composites. The obtained results displayed that the prepared nanocomposites displayed better tensile strength than that of foamed composites [39]. Spherical and layered Silica nanoparticles were melt blended with a polypropylene matrix. It was shown by the TEM images that the spherical NPs were dispersed in the polymer matrix.

3.2 Solution Blending Method

To disperse the Si fillers in the polymer matrix, solution blending is the best method [40]. By utilizing this method, nanofillers are effectively dispersed into the polymer chains. With a removal of the solvent, the well-dispersed Si filler/polymer hybrid composite can be achieved (Fig. 5). Generally, mesoporous Silica hybrid composites were synthesized by utilizing solution blending method [41–43]. For this process, the cyclic olefin copolymer is firstly suspended in THF and then the suspension is stirred at room temperature. This process is followed by drying at 70 °C under vacuum to form a polymer/Silica composite film [41]. The polyimide/Silica nanocomposite films derived from the solution blending process displayed good thermal stability and tensile modulus with enhanced Silica loading content [42]. An enhanced proton conductivity was displayed by the Nafion/mesoporous Silica composite synthesized in DMF. This was higher than that of the pure Nafion [44]. Yabu et al. reported the co-precipitation method for the synthesis of the composites of the SiNPs and polymer, for instances, poly (butadiene) in the THF [45].

3.3 In Situ Polymerization Method

In this method, the inorganic fillers are suspended in the liquid phase of the monomer and then the instigation of the polymerization takes place around and among the filler particles. The factors by which the reaction can be initiated include

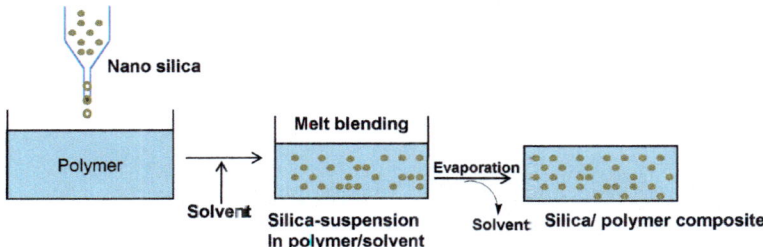

Fig. 5 Solution blending method

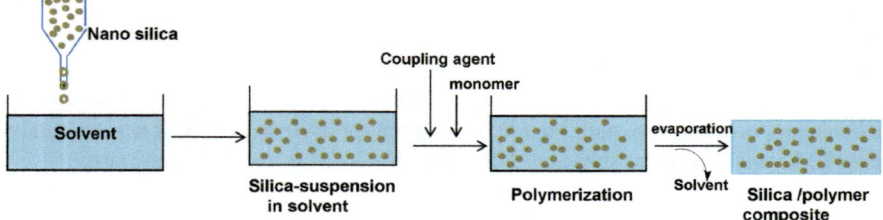

Fig. 6 In situ polymerization method

the curing agent, thermal activation, UV irradiation, incorporation of initiator, and the enzymatic initiation (Fig. 6). For the synthesis of the polymer grafting on Si, particles irradiation is the best route. It was demonstrated by the Kim et al. that the methyl methacrylate could be utilized for the surface alteration of the Silica nanofillers. This procedure was done by the UV-induced graft polymerization [46]. During the formation of the grafted Si nanofillers and polymers, the interface features depend on the types of the grafting polymers, which can be tunable for the microstructures and physical properties of the composites. When the p-vinyl phenyl sulfonyl hydrazide was grafted onto the nanosilica filler surface, an imperative role is played by the grafted agents in a blending with PP. This improves the compatibility of Silica with polymer matrix [47]. Modification of the Silica fillers with the help of g-methacryloxypropyl trimethoxysilane (MPS) was done during the preparation of nanostructural composites. These MPS-treated nanosilica fillers were utilized for in situ emulsion polymerization of butyl acrylate latex–Silica nanocomposite [48]. After the addition of these nanofillers, the mechanical properties of acrylate latex composites were considerably increased [48].

Core-shell hybrid composites of polyacrylamide layer on the Silica filler surface were prepared by Kohri et al. This process occurs in two steps: Firstly, the b-diketone structure having a silane coupling agent was immobilized on the Silica particle surface and then utilized as a polymerization initiator for horseradish peroxidase-mediated polymerization [49]. The interaction between Silica fillers and polymers obtained via the sol–gel method has been demonstrated in the solution states suspended with Silica particles. With the help of in situ bulk polymerization of a silicate sol and MMA monomers, hybrid materials based on silicate sol and polymethyl methacrylate were synthesized [50].

4 Applications

The Silica hybrid nanocomposites have unique physical properties which paying attention toward many industrial applications like common plastics, rubber reinforcement, and many other potentials and practical applications of Silica hybrid nanocomposite have been reported:

- Biomedical materials [51–56],
- Coating [57–59],
- Flame-retardant materials [60],
- Electronics and optical packaging materials [61, 62],
- Grouting materials [63, 64],
- Materials for metal uptake [65–69],
- Proton exchange membranes [70–74],
- Photoresist materials [75–77],
- Photoluminescent conducting film [78],
- Pervaporation membrane [79, 80],
- Oil adsorbents [81, 82],
- Optical devices [83],
- Sensors [84, 85],
- Ultra-permeable reverse-selective membranes [86].

4.1 Biomedical Application

Silica/polymer composites have attracted a noteworthy interest in various application fields, because they not only improve their physical properties such as the thermal, mechanical properties, but also exhibit some inimitable properties for instance optical transparency, weathering resistance, specific electrical, and abrasion resistance. The hybrid of silane forms with the modified amphiphilic chitosan (CS) by polymerizing with 3-aminopropyltriethoxysilane in the presence of carboxymethyl-hexanoyl chitosan and carbodiimide methiodide. The hybrid formed shows a stable polygonal geometry with 6 nm thick silane layers. This hybrid having good cytocompatibility shows a well-controlled encapsulation and releases profiles of (S)-(C)-camptothecin [87].

The hybrid of silicon oxide with carboxymethyl cellulose and silver (SiO_2-CMC-Ag) has excellent antibacterial and antimicrobial properties. The antimicrobial activity was tested with *E. coli* and *B. subtilis* in batch experiments with this Silica hybrid of SiO_2-CMC-Ag. The probable mechanism of interaction between organic and inorganic components involved the formation of new hydrogen bonds. In hybrids, Ag ions were present in valence states and in different size ratios. The hybrid also showed antibacterial activity against both Gram-positive and Gram-negative bacteria. The hybrid material showed antibacterial effect against Bacillus subtilis and Escherichia coli, respectively. The synthesis of the hybrid was carried out by utilizing CMC and silver [88, 89].

A new hybrid of HP-coated core-shell colloidal mesoporous Silica (CMS) NPs was utilized as an anticoagulant drug for the incorporation of biological functional molecules [90]. The Silica-coated magnetic particles may be utilized in engineering and biomedicine areas. These were negatively charged round, and superparamagnetic [91].

4.2 Coatings

Nowadays, researchers paid attention to the new class of coating having organic/inorganic hybrid materials. These coatings combine the flexibility and easy processing of polymers with the hardness of inorganic materials and have been successfully applied on various substrates. Generally, these hybrid coatings are having good adhesion, the enhanced property of scratch and abrasion resistance of polymer and transparent coatings [92]. The strengthening of acrylates coating material with the surface-improved nanosilica leads to better scratch and abrasion resistance coatings. These coatings can be used on substrates such as engineered wood, metals, paper, polymer films, and wood [93]. In comparison with the other nanocomposite coating materials, Silica-modified nanoparticle and corundum microparticles hybrids have much better abrasion resistance. The nano/micro Silica hybrid composites are suggested as clear coats for flooring and flooring coating applications [94]. The Silica nanocomposite hybrids with biopolymers were synthesized in different compositions and structures by utilizing mini-emulsion polymerization [95]. The resulting hybrid structures are utilized to produce water borne hybrid coatings.

4.3 Chemosensors

Polydiacetylene/Silica hybrid nanocomposite is utilized as chemosensors. The SiNPs adsorbed the aggregates of disordered 10,12-pentacosadiynoic acid (PCDA) in aqueous solution. The Si NPs template helps to make the disordered PCDA molecules in aggregates into an ordered PCDA arrangement. When the hybrid composite is irradiated with UV light, it seems to be a blue color. The environmental changes in temperature and pH could affect the result of a colorimetric change of hybrid during irradiation from blue to red phase. These hybrids have many attractive applications as new chemosensors [96].

4.4 Enzyme Immobilization (EI) and Sensor Application

The Silica–polymer hybrid composites have shown potential applications in sensor enlargement and enzyme immobilization. These types of hybrids provide the necessary conditions for the enzyme immobilization with the good catalytic activity of long-term preservation. These hybrids have advantages for the EI in respect for the enhanced stability, high adsorption property, and mechanical resistance of the immobilized enzymes.

A new silicate hybrid film at the anodized platinum electrode is synthesized by the sol–gel method by the Matsuhissa et al. for the glucose biosensor. These are efficient amperometric sensors having fast and sensitive to the glucose sensing in the presence of uric acid and ascorbic acid [97]. Further, a new generation Silica matrix with the incorporation of CNTs using carboxymethyl CS modified with pendant pyrene moieties by the sol–gel method is used for the electrochemical sensing. This hybrid has excellent stabilizing and dispersing role due to which MWCNTs were easily dispersed in the gel matrix and in the aqueous system. This type of gel has enhanced mechanical properties and shear thinning behavior. So, this type of gel was easily coated on the surface of the electrodes and to develop the biosensor [98, 99]. The Co-B-SiO$_2$ NPs hybrids are utilized for the immobilization of Go enzyme. The NPs entrapped beads (Co-BSiO$_2$-NH$_2$ NPs) were prepared by Cobalt-B-silicon oxide NPs that were derived by Stober polymerization of TEOS in the presence of Co-B. The immobilization of the enzyme increases the affinity, but it decreases the reactivity. The immobilization also increased the storage and operational stability of the enzyme. Liu et al. designed lanthanide doped CS-silica hybrid spheres with photoluminescent and local sensing features. The preparation of the hybrid spheres involved the synthesis of CS alcogel beads which were formed by dropping acetic acid solution of CS into aqueous NaOH solution. The partially hydrolyzed alkoxysilane species could penetrate and polymerize inside the core of these alcogel spheres. Then the spheres dried in the supercritical condition of carbon dioxide to maintain the porous texture than the dried atmosphere of CS was placed in a solution containing hydrogen peroxide, sodium fluoride, and tetraethyl silicate under vigorous stirring for 12 h. Finally, a homogeneous composite of the hybrid beads obtained at the supercritical. The obtained composite has core-shell type morphology and the thickness of the shell can be changed by changing the reaction time. Theses sphere hybrid bead is utilized in many biological applications like in the therapeutics and in the diagnostics. The Eu^{3+}containing hybrid showed red emission, while the Tb^{3+} containing hybrids, green emissions [50, 100]. Singh and Ahmed reported carboxymethyl cellulose (CMC)-silver nanoparticle (AgNPs)-silica hybrids for immobilizing alpha amylase. Hybrid gels were synthesized by polymerizing tetramethoxysilane (TMOS) in the presence of aqueous CMC solution that precontained AgNPs [101]. The hybrid adsorbed the enzyme, and the IE was utilized for optimization of soluble starch hydrolysis. The immobilization of enzyme increases the catalytic property of the free enzyme; this can be characterized by the kinetic parameters. The enzyme which is immobilized could be effortlessly stored and recycled.

An efficient hybrid of GA-gelatin has been utilized for the amylase immobilization. An efficient hybrid of GA-gelatin has been utilized for the amylase immobilization by GA–gelatin dual templated polymerization of TMOS in presence of AgNPs [102]. There is a no linkage between the inorganic and organic phase of the composite. 100 mg of the hybrid adsorbed the 40 mg of the enzyme for the immobilization. The amalgamation of silver nanoparticles increases the shelf life of the enzyme which is adsorbed but did not increase the bioactivity of the loaded enzyme. The IE shows excellent activity, stability, and the affinity as compared to the free enzyme in the solution.

TMSO polymerized in the presence of CMC and gelatin to obtain an efficient hybrid carrier support for the amylase. The most favorable activity shown by the nanohybrid could be obtained by using 10 mg gelatin, 200 mg CMC, 20 mL water, 1.5 mL TMOS, and 1.5 mL methanol. Then the obtained dried hydrogel was calcined at 300 °C to obtain hybrid xerogel. 100 mg of the presynthesized hybrid xerogel was adsorbed the 40 mg diastase alpha amylase. The immobilization did not change the most favorable condition of the enzymatic reaction such as the pH and the temperature. The immobilization is also enhanced by the catalytic function of alpha amylase. The IE has a higher potential of storage stability in comparison to the enzyme in solution [103].

4.5 Metal Uptake

The nanohybrid composites of Silica derivative, i.e., Silica oxide particles with electroactive polymers PANI or PPy have broad applications for heavy metal uptake. The metal uptake by these hybrids is based on the fact that they possess a surface area substantially higher than that estimated from the particle size and hence can aid the process of metal uptake. The use of electroactive polymer/SiO_2 nanocomposites for the uptake of gold and palladium from $AuCl_3$ and $PdCl_2$ in acid solutions, respectively was investigated. In the case of gold uptake, the reaction rate increased with temperature from 0 to 60 °C. The assembly of atomic Au on the nanocomposites increased the diameter and decreased the surface area. The uptake of Pd from $PdCl_2$ was much more complicated to achieve. To increase the rate of metal uptake, we must decrease the oxidation state and utilized the electro active polymers [104].

4.6 Proton Exchange Membranes (PEM)

The PEM is the important part of the solid fuel cells, for example, in the direct methanol fuel cells (DMFC) and in the proton exchange membrane fuel cell (PEMFC). Silica NPs form hybrids for proton exchange membranes, e.g., suffocated poly (phthalazinone ether ketone) (sPPEK) was mixed with Silica nanoparticles and forms PEMs [105]. This type of Silica hybrid membranes has enhanced properties of thermal stability, swelling behavior, and mechanical properties. Similarly, Silica hybrid membrane of sulfonated P(St-co-MA)-PEG/Si nanohybrid composite polyelectrolyte is used in the PEGS. The Silica content in the hybrid varies from using the different molecular weight of PEG which is responsible for the fine control over spacing between Silica domains [106]. These types of proton exchange membranes of Si hybrid have extensive applications in the DMFC. Although these membranes have no significant characteristics as compared to the Nation 117 membrane, Nation 117 membrane having good proton conductivity and as much as ion-exchange capacity.

Table 1 Recent Silica/polymer hybrid nanohybrid composites: preparations and properties

Type of Silica	Polymer	Blending methods	Specific properties	References
Aniline-modified Si particles	Polyaniline	Sol–gel	Good TS and Good mechanical strength	[108]
Nano Si	Silane coupling agents	Multi-axial aramid fabrics	Bullet shock impact for body armor	[109]
SiNPs	Polymers	Coatings and film	High tensile strength and impact resistance	[110]
Microsphere Si	Organophosphate polystyrene	Composition	Sorbents for Au	[111]
Microsphere Si	Poly[styrene-co-N-(4-vinylbenzyl)-N,N-diethylamine	3-(Methacryloxy) propyltrimethoxysilane	Sorbents for Pb ions	[112]
Hexagonal MS-Si	Polyaniline/polypyrrole	Composition	Sorbents for Cd	[104]
Si gels	Polyamidoamine grafted	Composition	Sorbents for Au	[113]
MSSi	Cellulose acetate	Post-spinning infusion tech	CO_2 capture	[114]
GO-Si	–	Hybrid coating	Hollow fiber solid phase and microextraction	[115]
MS Si	Polyacrylic acid	Gelation	pH responsive	[116]
Colloidal Si	Polycarbonate	GPTMOS cross-linking	Antifogging (AF) coating	[117]
SiNPs	Fullerene into the grafted poly (MMA-NMA)	Radical polymerization	Transparent C60/Silica	[118]
Cholinesterase immobilized Silica	Polysiloxane matrixes	Sol–gel synthesis	Cholinesterase activity	[119]
Amine-SiNPs encapsulated in Ethylene blue		Nanocomposite system	High p53 transfection—imaging and tumor-targeting in cancer	[120]
Si-coated magnetic GO	Polyglycerol-g-polycaprolactone		pH-responsive	[121]
Nanostructured porous silicon	Poly(L-lactide)	Composition	Controlled drug delivery	[122]
Epoxy/Si		Ag nanowire hybrid	High thermal conductivity	[123]

4.7 Evaporation Membranes

In this pervaporation separation method, the solution is placed in direct contact with the feed region membrane and waste or permeated is removed through vaporization from the region of the membrane. The transportation process of the solution throughout the membranes is different from other membrane processes like gas separation because the permeant in pervaporation frequently demonstrates high solubility in polymeric membranes. The Si hybrids or silane-modified Si fillers effect the pervaporation properties. The pervaporation process of PPO dense membranes has been reported. The methanol and methyl *tert*-butyl ether mixture were separated through pervaporation separation over the whole range of concentration. The separation process was carried out by utilizing both filled and unfilled membranes. The filled PPO membranes have high methanol selectivity and less permeability as compared to the separation of the unfilled PPO membrane. The modified SiNPs had higher affinity and enhanced compatibility with the PPO polymer than the unchanged SiNPs [107]. Recent Silica/polymer hybrid nanohybrid composites: preparations and properties are given in Table 1.

5 Conclusion

Many Silica/polymer composite materials have recently developed in various application fields due to their composites' unique properties combining the advantages of the inorganic fillers and the organic polymers. Owing to the combined advantages of the organic polymers and inorganic fillers, numerous Silica/polymer composite materials have been recently developed in diverse application fields. In the development of this silica polymers hybrid composites the overcome of various drawbacks, for instance, nanosized level hybridization, the inappropriateness of Silica with an organic polymer, and understanding to the interface interaction become critical issues. This review highlights the synthesis of Silica polymer nanocomposites as well as their applications. Even though bunches of research for the advancement of polymer/Silica nanocomposites have been reported in various fields, more review is required to additionally investigate the polymer lattice/filler composite structure and the relationship between hybrid structure and property.

Acknowledgements We gratefully acknowledge support from the Ministry of Science and Technology and Department of Science and Technology, Government of India under the scheme of Establishment of Women Technology Park, for providing the necessary financial support to carry out this study vide letter No, F. No SEED/WTP/063/2014.

Conflict of Interest The Authors do not have any conflict of interest.

References

1. Zou, H., S. Wu, and J. Shen. 2008. Polymer/silica nanocomposites: Preparation, characterization, properties, and applications. *Chemical Reviews* 108: 3893–3957. doi:10.1021/cr068035q.
2. Lee, D.W., and B.R. Yeo. 2014. Polymer nanotechnology: Nanocomposites. *Journal of Industrial and Engineering Chemistry* 20: 3204–3947.
3. Ciriminna, R., A. Fidalgo. V. Pandarus, F. Beland, L.M. Ilharco, and M. Pagliaro. 2013. The sol-gel route to advanced silica-based materials and recent applications. *Chemical Reviews* 113: 6592–6620. doi:10.1021/cr300399c.
4. Bergna H.E., and W.O. Roberts. 2015. *Colloidal silica: Fundamentals and applications*. CRC Press.
5. Iler, K. 1979. *The chemistry of silica: Solubility, polymerization, colloid and surface properties and biochemistry of silica*. Wiley.
6. Lee, M.S., and M.J. Jo. 2002. Coating of methyltriethoxysilane—Modified colloidal silica on polymer substrates for abrasion resistance. *Journal of Sol-Gel Science and Technology* 24: 175–180. doi:10.1023/A:1015208328256.
7. Althues, H., J. Henle, and S. Kaskel. 2007. Functional inorganic nanofillers for transparent polymers. *Chemical Society Reviews* 36: 1454–1465. doi:10.1039/B608177K.
8. Rambo, M.K.D., A.L. Cardoso, D.B. Bevilaqua, T.M. Rizzetti, L.A. Ramos, G.H. Korndorfer, and A.F. Martins. 2011. Silica from rice husk ash as an additive for rice plant. *Journal of Agronomy* 10: 99–104. doi:10.3923/ja.2011.99.104.
9. Shim, J., P. Velmurugan, and B.T. Oh. 2015. Extraction and physical characterization of amorphous silica made from corn cob ash at variable pH conditions via sol-gel processing. *Journal of Industrial and Engineering Chemistry* 30: 249–253. doi:10.1016/j.jiec.2015.05.029.
10. Velmurugan, P., J. Shim, K.J. Lee, S.S. Min Cho, S.K. Lim, K.M. Seo, K.S. Cho, B. Bang, and B.T. Oh. 2015. Extraction, characterization, and catalytic potential of amorphous silica from corn cobs by sol-gel method. *Journal of Industrial and Engineering Chemistry* 29: 298–303. doi:10.1016/j.jiec.2015.04.009.
11. Kellar, J.J. 2006. *Functional fillers and nanoscale minerals: New Markets/new Horisons*. SME.
12. Trewyn, B.G., I.I. Slowing, S. Giri, H.T. Chen, and V.S.Y. Lin. 2007. Synthesis and functionalization of a mesoporous silica nanoparticle-based on the sol-gel process and applications in controlled release. *Accounts of Chemical Research* 40: 846–853. doi:10.1021/ar600032u.
13. Zhao, B., and W. Brittain. 2000. Polymer brushes: Surface-immobilized macromolecules. *Progress in Polymer Science* 25: 677. doi:10.1016/S0079-6700(00),00012-5.
14. Belder, G.F., G.T. Brinke, and G. Hadziioannou. 1997. Influence of anchor block size on the thickness of adsorbed block copolymer layers. *Langmuir* 13: 4102–4105. doi:10.1021/la960379w.
15. Li, S., M.M. Lin, M.S. Toprak, D.K. Kim, and M. Muhammed. 2014. Nanocomposites of polymer and inorganic nanoparticles for optical and magnetic applications. *Nano Reviews* 1: 5214. doi:10.3402/nano.v1i0.5214.
16. Wei, L., N. Hu, and Y. Zhang. 2010. Synthesis of polymer—Mesoporous silica nanocomposites. *Materials* 3: 4066–4079. doi:10.3390/ma3074066.
17. Guyard, A., J. Persello, J.P. Boisvert, and B. Cabane. 2006. Relationship between the polymer/silica interaction and properties of silica composite materials. *Journal of Polymer Science Part B: Polymer Physics* 44: 1134. doi:10.1002/polb.20768.
18. Ahn, S.H., S.H. Kim, and S.G. Lee. 2004. Surface-modified silica nanoparticle–reinforced poly(ethylene 2,6-naphthalate). *Journal of Applied Polymer Science* 94: 812–818. doi:10.1002/app.21007.

19. Mahdavian, A.R., M. Ashjari, and A.B. Makoo. 2007. Preparation of poly (styrene–methyl methacrylate)/SiO$_2$ composite nanoparticles via emulsion polymerization. An investigation into the compatibilization. *European Polymer Journal* 43: 336–344. doi:10.1016/j.eurpolymj. 2006.10.004.
20. Tang, J.C., G.L. Lin, H.C. Yang, G.J. Jiang, and Y.W. Chen-Yang. 2007. Polyimide-silica nanocomposites exhibiting low thermal expansion coefficient and water absorption from surface-modified silica. *Journal of Applied Polymer Science* 104: 4096–4105. doi:10.1002/app.26041.
21. Ding, X.F., Z.C. Wang, D.X. Han, Y.J. Zhang, Y.F. Shen, Z.J. Wang, and L. Niu. 2006. An effective approach to the synthesis of poly(methyl methacrylate)/silica nanocomposites. *Nanotechnology* 17: 4796–4801. doi:10.1088/0957-4484/17/19/002.
22. Wu, T.M., and M.S. Chu. 2005. Preparation and characterization of thermoplastic vulcanizate/silica nanocomposites. *Journal of Applied Polymer Science* 98: 2058–2063. doi:10.1002/app.22406.
23. Lai, Y.H., M.C. Kuo, J.C. Huang, and M. Chen. 2007. On the PEEK composites reinforced by surface-modified nano-silica. *Materials Science and Engineering A* 458: 158–169. doi:10.1016/j.msea.2007.01.085.
24. Perro, A., S. Reculusa, E. Bourgeat-Lami, E. Duguet, and S. Ravaine. 2006. Synthesis of hybrid colloidal particles: From snowman-like to raspberry-like morphologies. *Colloids and Surfaces A: Physicochemical and Engineering Aspects* 284: 78–83. doi:10.1016/j.colsurfa. 2005.11.073.
25. Lin, J., J.A. Siddiqui, and R.M. Ottenbrite. 2001. Surface modification of inorganic oxide particles with silane coupling agent and organic dyes. *Polymers for Advanced Technologies* 12: 285–292. doi:10.1002/pat.64.
26. Kango, S., S. Kalia, A. Celli, J. Njuguna, Y. Habibi, and R. Kumar. 2013. Surface modification of inorganic nanoparticles for the development of organic–inorganic nanocomposites: A review. *Progress in Polymer Science* 38: 1232–1261. doi:10.1016/j.progpolymsci. 2013.02.003.
27. Plueddemann E.P. 1991. *Silane coupling agents*. 2nd ed. http://dx.doi.org/10.1007/978-1-4899-2070-6.
28. Yoo, B.R., D.E. Jung, and J.S. Han. 2009. *Materials Research Society Symposium Proceedings*, 1174-V06-08.
29. Bracho, D., V.N. Dougnac, H. Palza, and R. Quijada. 2012. Functionalization of silica nanoparticles for polypropylene nanocomposite applications. *Journal of Nanomaterials*, 263915. http://dx.doi.org/10.1155/2012/263915.
30. Zheng, K., L. Chen, Y. Li, and P. Cui. 2004. Preparation and thermal properties of silica-graft acrylonitrile-butadiene-styrene nanocomposites. *Polymer Engineering & Science* 44: 1077–1082. doi:10.1002/pen.20100.
31. Li, Chunzhao, and B.C. Benicewicz. 2005. Synthesis of well-defined polymer brushes grafted onto silica nanoparticles via surface reversible addition–fragmentation chain transfer polymerization. *Macromolecules* 38: 5929–5936. doi:10.1021/ma050216r.
32. Liu, G., M. Cai, F. Zhou, and W. Liu. 2014. Charged polymer brushes-grafted hollow silica nanoparticles as a novel promising material for simultaneous joint lubrication and treatment. *The Journal of Physical Chemistry B* 118: 4920–4931. doi:10.1021/jp500074g.
33. Zhang, M.Q., M.Z. Rong, and K. Friedrich. 2003. In *Handbook of organic-inorganic hybrid materials and nanocomposites*, vol. 2, ed. H.S. Nalwa, 113. Stevenson Ranch, CA: American Scientific Publishers.
34. Lei, Y.D., Z.H. Tang, B.C. Guo, L.X. Zhu, and D.M. Jia. 2010. Synthesis of novel functional liquid and its application as a modifier in SBR/silica composites. *Express Polymer Letters* 4: 692–703. doi:10.3144/expresspolymlett.2010.84.
35. Perez, L.D., B.L. Lopez. 2012 Thermal characterization of SBR/NBR blends reinforced with a mesoporous silica. *Journal of Applied Polymer Science*, E328–E333. doi:10.1002/app. 35689.

36. Suzuki, N., S. Kiba, Y. Kamachi, N. Miyamoto, and Y. Yamauchi. 2011. Mesoporous silica as smart inorganic filler: Preparation of robust silicone rubber with low thermal expansion property. *Journal of Materials Chemistry* 21: 5338–5344. doi:10.1039/C0JM03767B.
37. Suzuki, N., S. Kiba, and Y. Yamauchi. 2011. Bimodal filler system consisting of mesoporous silica particles and silica nanoparticles toward efficient suppression of thermal expansion in silica/epoxy composites. *Journal of Materials Chemistry* 2: 14941–14947. doi:10.1039/C1JM12405F.
38. Suzuki, N., S. Kiba, and Y. Yamauchi. 2011. Low dielectric property of novel mesoporous silica/polymer composites using smart molecular caps: Theoretical calculation of air space encapsulated inside mesopores. *Microporous and Mesoporous Materials* 138: 123–131. doi:10.1016/j.micromeso.2010.09.020.
39. Hwang, S., and P.P. Hsu. 2013. Effects of silica particle size on the structure and properties of polypropylene/silica composites foams. *Journal of Industrial and Engineering Chemistry* 19: 1377–1383. doi:10.1016/j.jiec.2012.12.043.
40. Mittal, V. 2015. *Synthesis techniques for polymer nanocomposites*. Wiley.
41. Ou, C.F., and M.C. Hsu. 2007. Preparation and characterization of cyclo olefin copolymer (COC)/silica nanoparticle composites by solution blending *Journal of Polymer Research* 14: 373–378. doi:10.1007/s10965-007-9119-5.
42. Huang, J.W., Y.L. Wen, C.C. Kang, and M.Y. Yeh. 2007. Preparation of polyimide-silica nanocomposites from nanoscale colloidal silica. *Polymer Journal* 39: 654–658. doi:10.1295/polymj.PJ2006217.
43. Yang, F., and G.L. Nelson 2004. PMMA/silica nanocomposite studies: Synthesis and properties. *Journal of Applied Polymer Science* 91: 3844–3850. doi:10.1002/app.13573.
44. Jin, Y.G., S.Z. Qiao, L. Zhang, Smarta S. XuaZP, J.C.D. da Costa, and G.Q. Lu. 2008. Novel Nafion composite membranes with mesoporous silica nanospheres as inorganic fillers. *Journal of Power Sources* 185: 664–669. doi:10.1016/j.jpowsour.2008.08.094.
45. Yabu, H., H. Satoh, M. Kanahara, Y. Saito, and M. Shimomura. 2014. Spontaneous formation of silica–polymer composite particles by simple co-precipitation process. *Japanese Journal of Applied Physics* 53: 05FT02.
46. Kim, S., E. Kim, and W. Kim. 2005. Surface modification of silica nanoparticles by UV-induced graft polymerization of methyl methacrylate. *Journal of Colloid and Interface Science* 292: 93–98. doi:10.1016/j.jcis.2005.09.046.
47. Cai, L.F., X.B. Huang, M.Z. Rong, W.H. Ruan, and M.Q. Zhang. 2006. Effect of grafted polymeric foaming agent on the structure and properties of nano-silica/polypropylene composites. *Polymer* 47: 7043–7050. doi:10.1016/j.polymer.2006.08.016.
48. Hashhemi-Nasab, R., and S.M. Mirabedini. 2013. Effect of silica nanoparticles surface treatment on in situ polymerization of styrene–butyl acrylate latex. *Progress in Organic Coatings* 76: 1016–1023. doi:10.1016/j.porgcoat.2013.02.016.
49. Fukushima, H., M. Kohri, T. Kojima, T. Taniguchi, K. Saito, and T. Nakahira. 2012. Surface-initiated enzymatic vinyl polymerization: Synthesis of polymer-grafted silica particles using horseradish peroxidase as catalyst. *Polym. Chem* 3: 1123–1125. doi:10.1039/c2py20036h.
50. Fu, H.P., R.Y. Hong, Y.J. Zhang, H.Z. Li, B. Xu, Y. Zheng, and D.G. Wei. 2009. Preparation and properties investigation of PMMA/silica composites derived from silicic acid. *Polymers for Advanced Technologies* 20: 84–91. doi:10.1002/pat.1226.
51. Pourjavadi, A., Z.M. Tehrani. and S. Joka. 2015. Functionalized mesoporous silica-coated magnetic graphene oxide by polyglycerol-g-polycaprolactone with pH-responsive behavior: Designed for targeted and controlled doxorubicin delivery. *Journal of Industrial and Engineering Chemistry* 28: 45–53. doi:10.1016/j.jiec.2015.01.021.
52. Salernitano, E., and C. Migliaresi. 2003. Composite materials for biomedical applications: A review. *Journal of Applied Biomaterials & Biomechanics* 1: 3–18.
53. Steven, J.P., M.Y. Irani, K. Williams, and N.H. Voelcker. 2012. Controlled drug delivery from composites of nanostructured porous silicon and poly (L-lactide). *Nanomedicine* 7: 995. doi:10.2217/nnm.11.176.

54. Payentko, V., A. Matkovsky, and Y. Matrunchik. 2015. Composites of silica with immobilized cholinesterase incorporated into polymeric shell. *Nanoscale Research Letters* 10: 82. doi:10.1186/s11671-015-0808-4.
55. Rho, W.Y., H.M. Kim, S. Kyeong, Y.L. Kang, D.H. Kim, H. Kang, C. Jeong, D.E. Kim, Y. S. Lee, and B.H. Jun. 2014. Facile synthesis of monodispersed silica-coated magnetic nanoparticles. *Journal of Industrial and Engineering Chemistry* 20: 2646–2649. doi:10.1016/j.jiec.2013.12.014.
56. Wu, H., Y. Zhao, X. Mu, H. Wu, L. Chen, W. Liu, Y. Mu, J. Liu, and X. Wei. 2015. A silica-polymer composite nanosystem for tumor-targeted imaging and p53 gene therapy of lung cancer. *Journal of Biomaterials Science, Polymer Edition* 26: 384–400. doi:10.1080/09205063.2015.1012035.
57. Zhang, T., L. Zhang, and C. Li. 2011. Study of the preparation and properties of PBT/Epoxy/SiO$_2$ nanocomposites. *Journal of Macromolecular Science - Physics* 50: 967. doi:10.1080/00222348.2010.497112.
58. Liu, H., J. Xu, B. Guo, and X. He. 2014. Preparation and performance of silica/polypropylene composite separator for lithium ion batteries. *Journal Materials Science* 49: 6961. doi:10.1007/s10853-014-8401-2.
59. Raveh, M., L. Liu, and D. Mandler. 2013. Electrochemical co-deposition of conductive polymer-silica hybrid thin films. *Physical Chemistry Chemical Physics* 15: 10876. doi:10.1039/c3cp50457c.
60. Kashiwagi, K., A.B. Morgan, J.M. Antonucci, M.R. VanLandingham, R.H. Harris, W.H. Awad, and J.R. Shields. 2003. Thermal and flammability properties of a silica–poly (methylmethacrylate) nanocomposite. *Journal of Applied Polymer Science* 89: 2072–2078. doi:10.1002/app.12307.
61. Chen, C., Y. Tang, Y.S. Ye, Z. Xue, Y. Xue, X. Xie, and Y.W. Mai. 2014. High-performance epoxy/silica coated silver nanowire composites as under fill material for electronic packaging. *Composites Science and Technology* 105: 80–85. doi:10.1016/j.compscitech.2014.10.002.
62. Wong, C.P., and R.S. Bollampally. 1999. Thermal conductivity, elastic modulus, and coefficient of thermal expansion of polymer composites filled with ceramic particles for electronic packaging. *Journal of Applied Polymer Science* 74: 3396–3403. doi:10.1002/(SICI)1097-4628(19991227)74:14<3396.
63. Yeh, J.M., and K.C. Chang. 2014. Nanofillers a surface coating—A review. *Journal of Industrial and Engineering Chemistry* 20: 275–291.
64. Golestaneh, M., G. Amini, G.D. Najafpour, and M.A. Beygi. 2010. Evaluation of mechanical strength of epoxy polymer concrete with silica powder as filler. *World Applied Sciences Journal* 9: 216.
65. Yin, P., M. Xu, W. Liu, R. Qu, X. Liu, and Q. Xu. 2014. High efficient adsorption of gold ions onto the novel functional composite silica microspheres encapsulated by organophosphonated polystyrene. *Journal of Industrial and Engineering Chemistry* 20: 379–390. doi:10.1016/j.jiec.2013.04.032.
66. Qu, R., X. Ma, M. Wang, C. Sun, X. Sun, S. Sun, Y. Zhang, and P. Yin. 2014. Homogeneous preparation of polyamidoamine grafted silica gels and their adsorption properties as A^{u3+} adsorbents. *Journal of Industrial and Engineering Chemistry* 20: 4382–4392. doi:10.1016/j.jiec.2014.02.005.
67. Tang, J., J. Sun, J. Xu, and W. Li. 2014. Grafting of poly[styrene-co-N-(4-vinylbenzyl)-N, N-diethylamine] polymer film onto the surface of silica microspheres and their application as an effective sorbent for lead ions. *Journal of Applied Polymer Science* 131: 39973. doi:10.1002/app.39973.
68. Taha, A.A., Y.N. Wu, H. Wang, and F. Li. 2012. Preparation and application of functionalized cellulose acetate/silica composite nanofibrous membrane via electrospinning for Cr(VI) ion removal from aqueous solution. *Journal of Environmental Management* 112: 10–16. doi:10.1016/j.jenvman.2012.05.031.

69. Mishra, A.K., T. Kuila, D.Y. Kim, N.H. Kim, and J.H. Lee. 2012. Protic ionic liquid-functionalized mesoporous silica-based hybrid membranes for proton exchange membrane fuel cells. *Journal of Materials Chemistry* 22: 24366–24372. doi:10.1039/C2JM33288D.
70. Guzmán, C., A. Alvarez O.E. Herrera, R. Nava, J.L. Garcia, L.A. Godínez, L.G. Arriaga, and W. Mérida. 2011. Water transport in composite membranes containing silica: temperature and relative humidity effects. *International Journal of Electrochemical Science* 6: 4648–4666. doi:10.1039/C2JM33288D.
71. Jang, S.Y., and S.H. Han. 2015. Sulfonated poly SEPS/hydrophilic-SiO_2 composite membranes for polymer electrolyte membranes (PEMs). *Journal of Industrial and Engineering Chemistry* 23: 285–289. doi:10.1016/j.jiec.2014.08.030.
72. Wang, H., A.B. Holmberg, L. Huang, Z. Wang, A. Mitra, J.M. Norbeck, and Y. Yan. 2002. Nafion-bifunctional silica composite proton conductive membrane. *Journal of Materials Chemistry* 12: 834–837. doi:10.1039/B107498A.
73. Liu, H., C. Gong, J. Wang, X. Liu, H. Liu, F. Cheng, G. Wang, G. Zheng, C. Qin, and S. Wen. 2016. Chitosan/silica coated carbon nanotubes composite proton exchange membranes for fuel cell applications. *Carbohydrate Polymers* 136: 1379–1385. doi:10.1016/j.carbpol.2015.09.085.
74. Kim, D.J., M.J. Jo, and S.Y. Nam. 2015. A review of polymer–nanocomposite electrolyte membranes for fuel cell application. *Journal of Industrial and Engineering Chemistry* 21: 36. doi:10.1016/j.jiec.2014.04.030.
75. Jang, J.H., C.K. Ullal, M. Maldovan, T. Gorishnyy, S. Kooi, C. Koh, and E.L. Thomas. 2007. 3D micro- and nanostructures via interference lithography. *Advanced Functional Materials* 17: 3027. doi:10.1002/adfm.200700140.
76. Cho, J.-D., H.-T. Ju, Y.-S. Park, and J.-W. Hong. 2006. Kinetics of cationic photopolymerizations of UV-curable epoxy-based SU8-Negative photoresists with and without silica nanoparticles. *Macromolecular Materials and Engineering* 291: 1155–1163. doi:10.1002/mame.200600124.
77. Li, Xingwei, Xiang Li, and Gengchao Wang. 2006. Conducting poly-N-[5-(8 quinolinol) ylmethyl]aniline/nano-SiO_2 composite with fluorescence. *Materials Letters* 60: 3342–3345. doi:10.1016/j.matlet.2006.03.025.
78. Liu, Y.L., C.Y. Hsu, Y.H. Su, and J.Y. Lai. 2005. Chitosan-silica complex membranes from sulfonic acid functionalized silica nanoparticles for pervaporation dehydration of ethanol-water solutions. *Biomacromolecules* 6: 368–373. doi:10.1021/bm049531w.
79. Khayet, M., J.P.G. Villaluenga, J.L. Valentin, M.A. López-Manchado, J.I. Mengual, and B. Seoane. 2005. Filled poly(2,6-dimethyl-1,4-phenylene oxide) dense membranes by silica and silane modified silica nanoparticles: Characterization and application in pervaporation. *Polymer* 46: 9881. doi:10.1016/j.polymer.2005.07.081.
80. Cho, Y.K., E.J. Park, and Y.D. Kim. 2014. Removal of oil by gelation using hydrophobic silica nanoparticles. *Journal of Industrial and Engineering Chemistry* 20: 1231. doi:10.1016/j.jiec.2013.08.005.
81. Nguyen, S.T., J. Feng, S.K. Ng, J.P.W. Wong, V.B.C. Tan, and H.M. Duong. 2014. Advanced thermal insulation and absorption properties of recycled cellulose aerogels. *Colloids and Surfaces A: Physicochemical and Engineering Aspects* 445: 128. doi:10.1016/j.colsurfa.2014.01.015.
82. Battaglin, G., E. Cattaruzza, F. Gonella, R. Polloni, B.F. Scremin, G. Mattei, P. Mazzoldi, and C. Sada. 2004. Structural and optical properties of Cu: Silica nanocomposite films prepared by co-sputtering deposition. *Applied Surface Science* 226: 52–56. doi:10.1016/j.apsusc.2003.11.030.
83. Chang, C.C., and W.C. Chen. 2002. Synthesis and optical properties of polyimide-silica hybrid thin films. *Chemistry of Materials* 14: 4242–4248. doi:10.1021/cm0202310.
84. Yu, Y.Y., and W.C. Chen. 2003. Transparent organic–inorganic hybrid thin films prepared from acrylic polymer and aqueous monodispersed colloidal silica. *Materials Chemistry and Physics* 82: 388–395. doi:10.1016/S0254-0584(03),00259-1.

85. Jang, J., J. Ha, and B. Lim. 2006. Synthesis and characterization of monodisperse silica–polyaniline core–shell nanoparticles. *Chemical Communications* 1622–1624.
86. Hsiao, M.H., T.H. Tung, C.S. Hsiao, and D. Liu. 2012. Nano-hybrid carboxymethyl-hexanoyl CS modified with (3-Aminopropyl) triethoxysilane for camptothecin delivery. *Carbohydrate Polymers* 89: 632–639. doi:10.1016/j.carbpol.2012.03.066.
87. Rangelova, N., L. Aleksandrov, T. Angelova, N. Georgieva, and R. Muller. 2014. Preparation and characterization of SiO_2/CMC/Ag hybrids with antibacterial properties. *Carbohydrate Polymers* 101: 1166–1175. doi:10.1016/j.carbpol.2013.10.041.
88. Rangelova, N., N. Georgieva, K. Mileva, R. Yuryev, and R. Muller. 2012. Synthesis and antibacterial activity of SiO_2-CMC-Ag hybrid materials prepared by sol-gel. *Comptes Rendus de l'Academie Bulgare des Sciences* 65: 1057–1064. doi:10.1080/13102818.2014.944789.
89. Argyo, C., V. Cauda, H. Engelke, J. Radler, G. Bein, and T. Bein. 2012. Heparin-coated colloidal mesoporous silica nanoparticles efficiently bind to antithrombin as an anticoagulant drug-delivery system. *Chemistry-A European Journal* 18: 428–432. doi:10.1002/chem.201102926.
90. Joanna, L., S. Magdalena, S. Michal, K. Mariusz, R. Marek, T. Waldemar, S. Agnieszka, K. Gabriela, and N. Maria. 2014. Synthesis and characterization of the superparamagnetic iron oxide nanoparticles modified with cationic CS and coated with silica shell. *Journal of Alloys and Compounds* 586: 45–51. doi:10.1016/j.jallcom.2013.10.039.
91. Soloukhin, V.A., W. Posthumus, J.C.M. Brokken-Zijp, and J. Loos. 2002. With, mechanical properties of silica–(meth)acrylate hybrid coatings on polycarbonate substrate G. *Polymer* 43: 6169. doi:10.1016/S0032-3861(02)00542-6.
92. Bauer, F., and R. Mehnert. 2005. UV curable acrylate nanocomposites: Properties and applications. *Journal of Polymer Research* 12: 483. doi:10.1007/s10965-005-4339-z.
93. Bauer, F., H.J. Gläsel, U. Decker, H. Ernst, A. Freyer, E. Hartmann, V. Sauerland, and R. Mehnert. 2003. Trialkoxysilane grafting onto nanoparticles for the preparation of clear coat polyacrylate systems with excellent scratch performance. *Progress in Organic Coatings* 47: 147. doi:10.1016/S0300-9440(03)00117-6.
94. Tiarks, F., K. Landfester, and M. Antoinette. 2001. Preparation of polymeric nanocapsules by miniemulsion polymerization. *Langmuir* 17: 908–918. doi:10.1021/la001276n.
95. Su, Y.L. 2006. Preparation of polydiacetylene/silica nanocomposite for use as chemosensors. *Reactive & Functional Polymers* 66: 967. doi:10.1016/j.reactfunctpolym.2006.01.021.
96. Matsuhisaa, H., M. Tsuchiyaa, and Y. Hasebe. 2013. Protein and polysaccharide-composite sol–gel silicate film for an interference-free amperometric glucose biosensor. *Colloids and Surfaces B* 111: 523–529. doi:10.1016/j.colsurfb.2013.06.046.
97. Zhang, L.M., G.H. Wang, and Z. Xing. 2011. Polysaccharide-assisted incorporation of multiwalled into sol–gel for electrochemical sensing. *Journal of Materials Chemistry* 21: 4650–4656. doi:10.1039/C0JM03031G.
98. Lei, L., Z. Cao, Q. Xie, Y. Fu, Y. Tan, M. Ma, and S. Yao. 2011. One-pot electrodeposition of 3-aminopropyltriethoxysilane–CS hybrid gel film to immobilize glucose oxidase for biosensing. *Sensors and Actuators, B: Chemical* 157: 282–289. doi:10.1016/j.snb.2011.03.063.
99. Lee, H.U.K., Y.S. Song, Y.J. Suh, C. Park, and S.W. Kim. 2012. Synthesis and characterization of glucose oxidase–core/shell magnetic nanoparticle complexes into CS bead. *Journal of Molecular Catalysis B: Enzymatic* 81: 31–36. doi:10.1016/j.molcatb.2012.05.004.
100. Liu, F., L.D. Carlos, R.A.S. Ferreira, J. Rocha, M.C. Ferro, A. Tourrette, F. Quignard, and M. Robitzer. 2010. Synthesis, texture, and photoluminescence of lanthanide-containing CS-silica hybrids. *Journal of Physical Chemistry* 114: 77–83. doi:10.1021/jp908563d.
101. Singh, V., and S. Ahmed. 2012. Synthesis and characterization of carboxymethyl cellulose-silver nanoparticle (AgNP)-silica hybrid for amylase immobilization. *Cellulose* 19: 1759–1769. doi:10.1007/s10570-012-9749-6.

102. Singh, V., and S. Ahmed. 2012. Silver nanoparticle (AgNPs) doped gum acacia-gelatin-silica nanohybrid: An effective support for diastase immobilization. *International Journal of Biological Macromolecules* 50: 353–361. doi:10.1016/j.ijbiomac. 2011.12.017.
103. Singh, V., and S. Ahmad. 2014. Carboxymethyl cellulose-gelatin-silica nanohybrid: An efficient carrier matrix for alpha amylase. *International Journal of Biological Macromolecules* 67: 439–445. doi:10.1007/s10570-012-9749-6.
104. Neoh, K.G., K.K. Tan, P.L. Goh, S.W. Huang, E.T. Kang, and K.L. Tan. 1999. Electroactive polymer–SiO_2 nanocomposites for metal uptake. *Polymer* 40: 887. doi:10.1016/S0032-3861 (98)00297-3.
105. Su, Y.H., Y.L. Liu, Y.M. Sun, J.Y. Lai, M.D. Guiver, and Y. Gao. 2006. Using silica nanoparticles for modifying sulfonated poly(phthalazinone ether ketone) membrane for direct methanol fuel cell: A significant improvement on cell performance. *Journal of Power Sources* 155: 111. doi:10.1016/j.jpowsour.2005.03.233.
106. Saxena, A., B.P. Tripathi and V.K. Shahi. 2007. Sulfonated Poly (styrene-co-maleic anhydride)–poly(ethylene glycol)–silica nanocomposite polyelectrolyte membranes for fuel cell applications. *The Journal of Physical Chemistry B* 111: 2454–12461. doi:10.1021/ jp072244c.
107. Khayet, M., J.P.G. Villaluenga, J.L. Valentin, M.A. López-Manchado, J.I. Mengual, and B. Seoane. 2005. Filled poly(2,6-dimethyl-1,4-phenylene oxide) dense membranes by silica and silane modified silica nanoparticles: Characterization and application in pervaporation. *Polymer* 46: 9881. doi:10.1016/j.polymer.2005.07.081.
108. Chang, K.C., C.H. Hsu, C.W. Feng, Y.Y. Huang, J.M. Yeh, H.P. Wan, and W.C. Hung. 2014. Preparation and comparative properties of membranes based on PANI and three inorganic fillers. *Express Polymer Letters* 8: 207–218. doi:10.3144/expresspolymlett.2014. 24.
109. Obradović, V., D.B. Stojanović, R. Jančić-Heinemann, I Živković, V. Radojević, P.S. Uskoković, and R. Aleksić. 2014. Ballistic properties of hybrid thermoplastic composites with silica nanoparticles. *Journal of Engineered Fibers and Fabrics* 9: 97–107.
110. Ribeiro, T., C. Baleizão, and J.P.S. Farinh. 2014. Functional films from silica/polymer nanoparticles. *Materials* 7: 3881. doi:10.3390/ma7053881.
111. Bhowmick, A.K., and H. Stephens. 2000. *Handbook of elastomers*. 2nd ed, 610. CRC Press.
112. Qu, R., X. Ma, M. Wang, C Sun, X. Sun, S. Sun, Y. Zhang, and P. Yin. 2014. Homogeneous preparation of polyamidoamine grafted silica gels and their adsorption properties as Au^{3+} adsorbents. *Journal of Industrial and Engineering Chemistry* 20: 4382–4392. doi:10.1016/j.jiec.2014.02 005.
113. Tang, J., J. Sun, J. Xu, and W. Li. 2014. Grafting of poly[styrene-co-N-(4-vinylbenzyl)-N, N-diethylamine] polymer film onto the surface of silica microspheres and their application as an effective sorbent for lead ions. *Journal of Applied Polymer Science* 131: 39973. doi:10. 1002/app.39973.
114. Rezaei, F., R.P. Lively, Y. Labreche, G. Chen, Y. Fan, W.J. Koros, and C.W. Jones. 2013. Aminosilane-grafted polymer/silica hollow fiber adsorbents for CO_2 capture from flue gas. *ACS Applied Materials & Interfaces* 5: 3921. doi:10.1021/am400636c.
115. Su, S., B. Chen, M. He, and B. Hu. 2014. Graphene oxide–silica composite coating hollow fiber solid phase microextraction online coupled with inductively coupled plasma mass spectrometry for the determination of trace heavy metals in environmental water samples. *Talanta* 123: 1–9. doi:10.1016/j.talanta.2014.01.061.
116. Taha, A.A., Y.N. Wu, H Wang, and F. Li. 2012. Preparation and application of functionalized cellulose acetate/silica composite nanofibrous membrane via electrospinning for Cr (VI) ion removal from aqueous solution. *Journal of Environmental Management* 112: 10–16. doi:10.1016/j.jenvman.2012.05.031.
117. Chang, C.C., K.C. Wang, C.C. Chen, and L.P. Cheng. 2014. Preparation and characterization of silica/polymer antifogging coatings. *Polymers & Polymer Composites* 22: 39–44.

118. Yoshinaga, K., Y. Yang, T. Ohno, S. Motokucho, and K. Kojio. 2014. Inclusion of fullerene in polymer chains grafted on silica nanoparticles in an organic solvent. *Polymer Journal* 46: 623–627. doi:10.1038/pj.2014.24.
119. Payentko, V., A. Matkovsky, and Y. Matrunchik. 2015. Composites of silica with immobilized cholinesterase incorporated into polymeric shell Nanoscale Res. *Letters* 10: 82. doi:10.1186/s11671-015-0808-4.
120. Wu, H., Y. Zhao, X. Mu, H. Wu, L. Chen, W. Liu, Y. Mu, J. Liu, and X. Wei. 2015. A silica-polymer composite nano system for tumor-targeted imaging and p53 gene therapy of lung cancer. *Journal of Biomaterials Science, Polymer Edition* 26: 384–400. doi:10.1080/09205063.2015.1012035.
121. Samart, C., P. Prawingwong, S. Amnuaypanich, H. Zhang, K. Kajiyoshi, and P. Reubroycharoen. 2014. Preparation of poly acrylic acid grafted mesoporous silica as pH responsive releasing material. *Journal of Industrial and Engineering Chemistry* 20: 2153–2158. doi:10.1016/j.jiec.2013.09.045.
122. McInnes, Steven J.P., Y. Irani, K.A. Williams, and N.H. Voelcker. 2012. Controlled drug delivery from composites of nanostructured porous silicon and poly(L-lactide). *Nanomedicine* 7: 995–1016. doi:10.2217/nnm.11.176.
123. Gonon, P., A. Sylvestre, J. Teysseyre, and C. Prior. 2001. Dielectric properties of epoxy/silica composites used for microlectronic packaging, and their dependence on post-curing. *Journal Materials Science* 12: 81. doi:10.1023/A:1011241818209.

Printed by Printforce, the Netherlands